VOLUME EIGHTY ONE

ADVANCES IN
GENETICS

ADVANCES IN GENETICS, VOLUME 81

Serial Editors

Theodore Friedmann
University of California at San Diego,
School of Medicine, CA, USA

Jay C. Dunlap
The Geisel School of Medicine at Dartmouth,
Hanover, NH, USA

Stephen F. Goodwin
University of Oxford, Oxford, UK

VOLUME EIGHTY ONE

ADVANCES IN
GENETICS

Edited by

Theodore Friedmann
Department of Pediatrics,
University of California at San Diego,
School of Medicine, CA, USA

Jay C. Dunlap
Department of Genetics,
The Geisel School of Medicine at Dartmouth,
Hanover, NH, USA

Stephen F. Goodwin
Department of Physiology,
Anatomy and Genetics,
University of Oxford,
Oxford, UK

AMSTERDAM • BOSTON • HEIDELBERG • LONDON
NEW YORK • OXFORD • PARIS • SAN DIEGO
SAN FRANCISCO • SINGAPORE • SYDNEY • TOKYO

ELSEVIER   Academic Press is an imprint of Elsevier

Academic Press is an imprint of Elsevier
225 Wyman Street, Waltham, MA 02451, USA
525 B Street, Suite 1800, San Diego, CA 92101-4495, USA
Radarweg 29, PObox 211, 1000 AE Amsterdam, The Netherlands
The Boulevard, Langford Lane, Kidlington, Oxford, OX5 1GB, UK
32 Jamestown Road, London, NW1 7BY, UK

First edition 2013

ISBN: 978-0-12-407677-8
ISSN: 0065-2660

For information on all Academic Press publications
visit our website at store.elsevier.com

Printed and bound in USA
13 14 15   10 9 8 7 6 5 4 3 2 1

Working together to grow
libraries in developing countries

www.elsevier.com | www.bookaid.org | www.sabre.org

ELSEVIER   BOOK AID
            International   Sabre Foundation

CONTENTS

CONTRIBUTORS

Laura Almasy
Department of Genetics, Texas Biomedical Research Institute, San Antonio, Texas, USA

Marcio Almeida
Department of Genetics, Texas Biomedical Research Institute, San Antonio, Texas, USA

John Blangero
Department of Genetics, Texas Biomedical Research Institute, San Antonio, Texas, USA

J. Lesley Brown
Eunice Kennedy Shriver National Institute of Child Health and Human Development, National Institutes of Health, Bethesda, Maryland, USA

Vincent P. Diego
Department of Genetics, Texas Biomedical Research Institute, San Antonio, Texas, USA

Thomas D. Dyer
Department of Genetics, Texas Biomedical Research Institute, San Antonio, Texas, USA

Stephen J. Free
Department of Biological Sciences, SUNY, University at Buffalo, Buffalo, New York, USA

Harald H.H. Göring
Department of Genetics, Texas Biomedical Research Institute, San Antonio, Texas, USA

Judith A. Kassis
Eunice Kennedy Shriver National Institute of Child Health and Human Development, National Institutes of Health, Bethesda, Maryland, USA

Jack W. Kent Jr.
Department of Genetics, Texas Biomedical Research Institute, San Antonio, Texas, USA

Juan Peralta
Department of Genetics, Texas Biomedical Research Institute, San Antonio, Texas, USA

Jeff T. Williams
Department of Genetics, Texas Biomedical Research Institute, San Antonio, Texas, USA

A Kernel of Truth: Statistical Advances in Polygenic Variance Component Models for Complex Human Pedigrees

John Blangero[1], Vincent P. Diego, Thomas D. Dyer, Marcio Almeida, Juan Peralta, Jack W. Kent Jr., Jeff T. Williams, Laura Almasy, Harald H. Göring

Department of Genetics, Texas Biomedical Research Institute, San Antonio, Texas, USA
[1]Corresponding author: e-mail address: john@txbiomedgenetics.org

Contents

Abstract

Statistical genetic analysis of quantitative traits in large pedigrees is a formidable computational task due to the necessity of taking the nonindependence among relatives into account. With the growing awareness that rare sequence variants may be important in human quantitative variation, heritability and association study designs involving large pedigrees will increase in frequency due to the greater chance of observing multiple copies of rare variants among related individuals. Therefore, it is important to have statistical genetic test procedures that utilize all available information for extracting evidence regarding genetic association. Optimal testing for marker/phenotype association involves the exact calculation of the likelihood ratio statistic which requires the repeated inversion of potentially large matrices. In a whole genome sequence association context, such computation may be prohibitive. Toward this end, we have developed a rapid and efficient eigensimplification of the likelihood that makes analysis of family data commensurate with the analysis of a comparable sample of unrelated individuals. Our theoretical results which are based on a spectral representation of the likelihood yield simple exact expressions for the expected likelihood ratio test statistic (ELRT) for pedigrees of arbitrary size and complexity. For heritability, the ELRT is

$$-\sum \ln\left[1 + h^2\left(\lambda_{gi} - 1\right)\right],$$

where h^2 and λ_{gi} are, respectively, the heritability and eigenvalues of the pedigree-derived genetic relationship kernel (GRK). For association analysis of sequence variants, the ELRT is given by

$$\text{ELRT}\left[h_q^2 > 0 : \text{unrelateds}\right] - \left(\text{ELRT}\left[h_t^2 > 0 : \text{pedigrees}\right] - \text{ELRT}\left[h_r^2 > 0 : \text{pedigrees}\right]\right),$$

where h_t^2, h_q^2, and h_r^2 are the total, quantitative trait nucleotide, and residual heritabilities, respectively. Using these results, fast and accurate analytical power analyses are possible, eliminating the need for computer simulation. Additional benefits of eigensimplification include a simple method for calculation of the exact distribution of the ELRT under the null hypothesis which turns out to differ from that expected under the usual asymptotic theory. Further, when combined with the use of empirical GRKs—estimated over a large number of genetic markers—our theory reveals potential problems associated with non-positive semidefinite kernels. These procedures are being added to our general statistical genetic computer package, SOLAR.

1. INTRODUCTION

With the rise of next generation sequencing (NGS) and the resulting increase in available whole genome sequence (WGS), the modern statistical genetics of complex disease-related phenotypes finds itself confronted with the Herculean task of analyzing an astronomical volume of data. Of particular importance is the fact that, by far, most human sequence variation is rare (1000 Genomes Project Consortium et al., 2012), so rare that much sequence variation is effectively private (or lineage-specific). That fraction of the genome that we are most interested in, the phenotypically functional component, is even more likely to be dominated by such rare variation. In man, rare functional variation is best studied in large pedigrees. Basically, pedigree-based studies represent an implicit enrichment strategy for identifying and studying rare functional variants. Mendelian transmissions from parents to offspring maximize the chance that multiple copies of rare variants exist in the pedigree. Alternatively, studies of unrelated individuals like those typically performed in the now receding genome-wide association (GWA) era that have focused only upon common sequence variation can never capture more than one copy of a "private" variant. While there are accumulating methods to examine the joint effects of sequence variation in a gene-centric manner that may be of value in the study of unrelateds, a large part of human genetics will stay focused on the rapid identification of specific rare variants of moderate to large effect on disease risk since such variants more rapidly lead to functional experimental validation and causal gene discovery with all of its concomitant benefits. Thus, it is apparent that the coming WGS era of human genetics will require a return to our fundamental roots with a refocus on pedigree-based studies of phenotypic variation (Blangero, 2004; Ott, Kamatani, & Lathrop, 2011).

The analysis of the most valuable kinds (for studying rare variation) of large and complex human pedigrees has its own difficulties including substantial statistical and computational issues. At first glance, it would seem anachronistic to attack this issue by retreating to the classical methods of polygenic analysis under a variance component (VC) model, which have their origins almost a century ago now in Fisher (1918). This linear mixed model which allows for the simultaneous analysis of both fixed (e.g., the effects of specific sequence variants on the mean) and random effects (typically the residual polygenic effects and random environmental effects) has been successfully used for many years in human pedigree analysis. However, usage of VC models in large human pedigrees of the kind most likely to be valuable for the study of rare sequence variation has generally been computationally formidable. Similarly, obtaining accurate pedigree information itself is a difficult task in human populations (and especially isolated populations).

In this work, we demonstrate two advances in polygenic VC analysis that can be used to rationally analyze WGS variation in relation to its effects on phenotypic variation or disease risk. Specifically, we describe an eigenvalue decomposition (EVD) approach to likelihood analysis under a VC polygenic model (hereafter polygenic model) that greatly simplifies/speeds analyses and, more importantly, leads to a remarkable set of closed form analytical equations for power analyses for both heritability studies and marker-based association studies in arbitrary pedigrees. Additionally, this spectral decomposition of the likelihood function effectively removes all barriers to computation for even the largest and most complex of pedigrees. We also describe the use of empirical genetic relationship kernels (GRKs) that substantially broadens the potential to use polygenic models in the absence (or in support) of accurate pedigree information. We tie the two approaches together in a section where we use the EVD-derived likelihood approach to study the statistical properties of a typical GRK usage.

2. VCs MODELS

2.1. Standard polygenic model

We start with a standard description of the linear model for a phenotype vector under a VC model, which is a standard modeling approach for human family data (Almasy & Blangero, 1998, 2010; Blangero, Williams, & Almasy, 2001; Lange, 2002):

$$\mathbf{y}_{n\times1} = \mathbf{X}_{n\times j}\boldsymbol{\beta}_{j\times1} + \mathbf{g} + \mathbf{e}, \tag{1.1}$$

where $\mathbf{y}$, the phenotype vector of interest, $\mathbf{X}$, a design matrix of covariate effects, and $\boldsymbol{\beta}$, a vector of regression coefficients, are of dimensions $n\times1$, $n\times j$, and $j\times1$, respectively, and n and j give the numbers of individuals in the pedigree and of fixed-effect parameters, respectively, and $\mathbf{g}$ and $\mathbf{e}$ are unobserved vectors of random genetic and environmental effects, respectively. On assuming that the genetic and environmental effects are uncorrelated, the polygenic model for the phenotypic covariance matrix is as follows:

$$V[\mathbf{y}] = \boldsymbol{\Sigma} = \mathbf{K}h^2 + \mathbf{I}\left(1 - h^2\right), \tag{1.2}$$

where $\mathbf{K}$ is the GRK (which is also known as a genetic relationship matrix), $\mathbf{I}$ is the identity matrix, and $h^2 = \sigma_g^2/(\sigma_g^2 + \sigma_e^2) = \sigma_g^2/\sigma_p^2$ is the standard additive genetic heritability, where $\sigma_g^2, \sigma_e^2,$ and σ_p^2 are the additive genetic, residual environmental, and total phenotypic variances, respectively. For this basic model, $\mathbf{K} = 2\boldsymbol{\Phi}$, where $\boldsymbol{\Phi}$ is the expected kinship matrix generally derived directly from pedigree information. Assuming that the trait follows a multivariate normal (MVN) distribution, the model ln-likelihood function is given as

$$\ln L\left(\boldsymbol{\beta}, h^2 | \mathbf{y}, \mathbf{X}\right) = -\frac{1}{2}\left[N\ln 2\pi + \ln|\boldsymbol{\Sigma}| + \boldsymbol{\delta}'\boldsymbol{\Sigma}^{-1}\boldsymbol{\delta}\right], \tag{1.3}$$

where $\boldsymbol{\delta} = \mathbf{y} - \mathbf{X}\boldsymbol{\beta}$. If the data do not conform to the MVN assumption, we generally advocate direct inverse Gaussian transformation either prior to analysis or post initial covariate adjustment.

Following Boerwinkle, Chakraborty, and Sing (1986), a measured genotype (MG) effect at a single nucleotide polymorphism (SNP) may be included in the model for the mean as a fixed-effect parameter. Earlier approaches to incorporate an MG effect were made by Moll, Powsner, and Sing (1979) as a fixed effect and by Hopper and Matthews (1982) as a random effect, but the mature MG model was fully developed by Boerwinkle et al. (1986). Casual inspection of the likelihood equation (Equation 1.3) shows that likelihood analysis to the tune of one SNP at a time can be computationally burdensome for a large number of SNPs, and for large pedigrees. This is because computation of the inverse covariance matrix of the pedigree is required each time the likelihood is maximized in order to find the maximum likelihood estimates (MLEs). In the current GWA study designs employing large extended families and having total sample sizes of about a thousand (or appreciably more) individuals, and where a million SNPs are to be analyzed, we would have to invert the covariance matrix of size, say, 1000×1000 for at least $1\,M \times$ (number of likelihood evaluations) times. Obviously, this problem is amplified for NGS data analysis where the number of sequence variants to be analyzed can easily approach 25 M in a study of similar size.

2.2. Eigensimplification of the MVN likelihood

Because of the computational burden inherent in the traditional analytical approach, we earlier proposed a simplified approach to the problem using the EVD of the covariance matrix (Dyer, Diego, Kent, Göring, & Blanger, 2009). We call this general process the eigensimplification of the likelihood function. Hints or variations of the basic EVD method have been developed in relation to maximum likelihood estimation and have been applied in statistics and genetics for decades, always as a means of simplifying the attendant computational rigor (Dempster, Patel, Selwyn, & Roth, 1985; Kang et al., 2008, 2010; Patterson & Thompson, 1971; Thompson, 1973, 2008; Thompson & Cameron, 1986; Thompson & Meyer, 1986; Thompson & Shaw, 1990, 1992). Here, we similarly employ an orthogonal transformation of the data vector which maps or linearly transforms a vector of non-independent observations to a vector of independent observations. If the trait data were sampled from unrelated individuals, the likelihood would involve the simple product of univariate normal densities. However, because data sampled from families are inherently nonindependent, we must account for the nonindependence generated by genetic transmission. After the orthogonal transformation, we will see that the data are "decorrelated" or "whitened," which essentially diagonalizes the covariance matrix and reduces the likelihood again to the product of univariate normal densities. This consequence arises simply because the data vector has been taken from a vector space of nonindependent observations into a vector space of independent observations by way of an orthogonal transformation to the eigenbasis. Figure 1.1 represents a graphic depiction of this process where a bivariate probability density is transformed into two univariate probability densities.

Assuming for convenience and with complete generality that $\sigma_p^2 = 1$, the EVD of the covariance matrix can be written as:

$$\mathbf{\Sigma} = \mathbf{SD_pS'} = \mathbf{S}\left[h^2\mathbf{D_g} + \left(1 - h^2\right)\mathbf{I}\right]\mathbf{S'} = \mathbf{S}\left[\mathbf{I} + h^2\left(\mathbf{D_g} - \mathbf{I}\right)\right]\mathbf{S'}, \qquad (1.4)$$

where $\mathbf{S}$ is an orthogonal matrix of eigenvectors, and $\mathbf{D_p} = \text{diag}\{\lambda_{pi}\}$ and $\mathbf{D_g} = \text{diag}\{\lambda_{gi}\}$ are, respectively, diagonal matrices of phenotypic and additive genetic eigenvalues. The simple linear form for the phenotypic eigenvalues represents the critical component leading to dramatic speed-up of likelihood function evaluation. Likelihood computations can now be greatly simplified by employing a linear transformation of the vector of residuals to the eigenbasis of the covariance matrix, which we denote by $\mathbf{\tau}$:

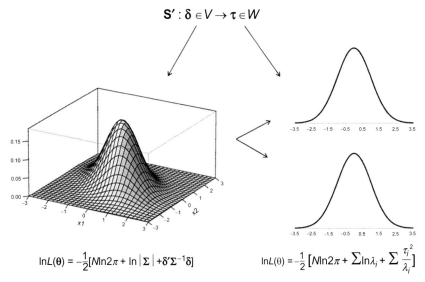

$$\mathbf{S'} : \delta \in V \rightarrow \tau \in W$$

$$\ln L(\theta) = -\frac{1}{2}[N\ln 2\pi + \ln|\mathbf{\Sigma}| + \delta'\mathbf{\Sigma}^{-1}\delta]$$

$$\ln L(\theta) = -\frac{1}{2}\left[N\ln 2\pi + \Sigma\ln\lambda_i + \Sigma\frac{\tau_i^2}{\lambda_i}\right]$$

Figure 1.1 Orthogonal transformation of the residuals vector. Schematic representation of the linear mapping of a vector in vector space V (of nonindependent data) to a vector in vector space W (of independent data).

$$\boldsymbol{\tau} = \mathbf{S}'\boldsymbol{\delta}. \tag{1.5}$$

Since $\boldsymbol{\delta}$ is MVN, so is the vector of transformed variables (Anderson, 1984):

$$\boldsymbol{\tau} \sim N(\mathbf{S}'\boldsymbol{\mu}, \mathbf{S}'\boldsymbol{\Sigma}\mathbf{S}),$$

where $V[\boldsymbol{\tau}] = \mathbf{S}'\boldsymbol{\Sigma}\mathbf{S} = \mathbf{I}$. One of the chief virtues of this approach, besides leading to a simplified likelihood, is that $\mathbf{S}$ and $\mathbf{D}_g$ can be computed from an initial EVD of $\mathbf{K}$ which needs to be performed only once before subsequent model evaluations. That the EVD of $\mathbf{K}$ is sufficient for our purposes is made possible by the facts that the eigenvectors of $\boldsymbol{\Sigma}$ are also the eigenvectors of $\mathbf{K}$ (Thompson & Shaw, 1990, 1992), and the eigenvalues of $\boldsymbol{\Sigma}$ can be written as a linear function of the eigenvalues of $\mathbf{K}$ in the manner stated above. From standard multivariate theory (Stuart & Ord, 1987), we know that the full likelihood is factored into the likelihood for the transformed trait and the likelihood for the transformation, where the latter is given by the Jacobian of the transformation, which is denoted by $J_{\boldsymbol{\delta}\rightarrow\boldsymbol{\tau}}$. Thus, the full likelihood will be on the natural logarithmic scale a sum of the likelihood of the transformed variable and the natural logarithm of the Jacobian of the transformation:

$$
\begin{aligned}
\ln L(\boldsymbol{\theta}|\mathbf{y},\mathbf{X}) &= \ln L(\boldsymbol{\tau}) + \ln(J_{\boldsymbol{\delta}\rightarrow\boldsymbol{\tau}}) \\
&= -\frac{1}{2}\Big[\ln|\mathbf{S}'\boldsymbol{\Sigma}\mathbf{S}| + \tau'(\mathbf{S}'\boldsymbol{\Sigma}\mathbf{S})^{-1}\boldsymbol{\tau}\Big] + \ln(J_{\boldsymbol{\delta}\rightarrow\boldsymbol{\tau}}) \\
&= -\frac{1}{2}\Big[\ln|\mathbf{S}'(\mathbf{S}\mathbf{D}_p\mathbf{S}')\mathbf{S}| + \boldsymbol{\tau}'(\mathbf{S}'(\mathbf{S}\mathbf{D}_p\mathbf{S}')\mathbf{S})^{-1}\boldsymbol{\tau}\Big] + \ln|\mathbf{S}'| \\
&= -\frac{1}{2}\Big[\ln|\mathbf{I}\mathbf{D}_p\mathbf{I}| + \boldsymbol{\tau}'(\mathbf{I}\mathbf{D}_p\mathbf{I})^{-1}\boldsymbol{\tau}\Big] \\
&= -\frac{1}{2}\Big[\ln|\mathbf{I} + h^2(\mathbf{D}_g - \mathbf{I})| + \boldsymbol{\tau}'[\mathbf{I} + h^2(\mathbf{D}_g - \mathbf{I})]^{-1}\boldsymbol{\tau}\Big] \\
&= -\frac{1}{2}\sum \ln\big[1 + h^2(\lambda_{gi} - 1)\big] - \frac{1}{2}\sum \frac{\tau_i^2}{1 + h^2(\lambda_{gi} - 1)},
\end{aligned}
\tag{1.6}
$$

where $\boldsymbol{\theta} = [\boldsymbol{\beta}, h^2]'$ is the parameter vector, $\mathbf{S}'\mathbf{S} = \mathbf{S}\mathbf{S}' = \mathbf{I}$ by definition of an orthogonal matrix, $|\mathbf{S}'| = 1$ on restricting $\mathbf{S}'$ to be a rotation (Abadir & Magnus, 2005; Pettofrezzo, 1978), and the summations are taken over n. Note that the phenotypic eigenvalues are reexpressed as a function of the heritability and the additive genetic eigenvalues. The major result of this spectral decomposition is that the likelihood has been simplified to be a sum of univariate likelihoods, as would be the case for the total likelihood for a sample of unrelated individuals or independent observations. It is interesting to observe that similar simplified likelihoods have been proposed under similar conditions involving the eigenvalues of the covariance matrix (Anderson & Olkin, 1985), but to our knowledge it appears that these simplified likelihoods were not utilized until only recently in statistical genetics in the context of the linear mixed model (Kang et al., 2008). Importantly, we note that because of the linear simplicity of Equation (1.4), the required spectral decomposition of the GRK needs to be done only once and the transformation can be performed on the phenotype and covariate vector prior to analysis. For real data, this will be even true across the evaluation of very different models as long as the missing data pattern (for both phenotype and covariates) is constant. This fact was neither noted nor implemented by Kang et al. (2008). Remarkably, our eigensimplification of the likelihood results in a rapid exact calculation of the usual MVN likelihood that is equivalent in speed to that observed for an equal number of unrelated subjects.

3. EXPECTED LIKELIHOOD RATIO TEST STATISTICS

3.1. Heritability

The eigensimplification of the multivariate likelihood in Equation (1.6) leads to some very useful analytical results of substantial relevance for the genetic analysis of phenotypic variation in arbitrary pedigrees. To show some of these results, it is convenient to work with the expected likelihood ratio test statistic, denoted as ELRT. We employ the ELRT for several reasons: (1) it is the easiest test statistic to analytically derive in comparison to asymptotically equivalent alternatives, (2) it provides an asymptotically uniformly most powerful test statistic for a VC, and (3) the ELRT leads to dramatically simplified analytical power and relative efficiency analyses.

To derive the ELRT, we require the following expectation:

$$E\big(\boldsymbol{\tau}'(\mathbf{S}'\boldsymbol{\Sigma}\mathbf{S})^{-1}\boldsymbol{\tau}\big) = (\mathbf{S}'\boldsymbol{\Sigma}\mathbf{S})^{-1}E(\boldsymbol{\tau}\boldsymbol{\tau}') = (\mathbf{S}'\boldsymbol{\Sigma}\mathbf{S})^{-1}E\big(\mathbf{S}'\boldsymbol{\delta}(\mathbf{S}'\boldsymbol{\delta})'\big)$$
$$= (\mathbf{S}'\boldsymbol{\Sigma}\mathbf{S})^{-1}\mathbf{S}'E(\boldsymbol{\delta}\boldsymbol{\delta}')\mathbf{S} = (\mathbf{S}'\boldsymbol{\Sigma}\mathbf{S})^{-1}\mathbf{S}'\boldsymbol{\Sigma}\mathbf{S} = \mathbf{I},$$

which shows that the quadratic term in the likelihood function cancels out on taking the difference in the ELRT. Thus, for a test of total additive genetic heritability, we find

$$\text{ELRT}[h^2 > 0] = E\left[-2\left(\ln L\left(h_0^2 = 0\right) - \ln L\left(h^2\right)\right)\right]$$

$$= E\left[-2\left(-\frac{1}{2}\left[\ln|\mathbf{S}'\boldsymbol{\Sigma}_\text{N}\mathbf{S}| + \boldsymbol{\tau}'(\mathbf{S}'\boldsymbol{\Sigma}_\text{N}\mathbf{S})^{-1}\boldsymbol{\tau}\right]\right.\right.$$

$$\left.\left.+\frac{1}{2}\left[\ln|\mathbf{S}'\boldsymbol{\Sigma}_\text{A}\mathbf{S}| + \boldsymbol{\tau}'(\mathbf{S}'\boldsymbol{\Sigma}_\text{A}\mathbf{S})^{-1}\boldsymbol{\tau}\right]\right)\right]$$

$$= \left(\ln|\mathbf{D}_\text{pN}| + \mathbf{I}\right) - \left(\ln|\mathbf{D}_\text{pA}| + \mathbf{I}\right)$$

$$= \ln|\mathbf{I} + h_0^2\left(\mathbf{D}_\text{g} - \mathbf{I}\right)| - \ln|\mathbf{I} + h^2\left(\mathbf{D}_\text{g} - \mathbf{I}\right)|$$

$$= -\sum \ln\left[1 + h^2\left(\lambda_\text{gi} - 1\right)\right], \tag{1.7}$$

where the covariance matrices, and diagonal matrices of phenotypic eigen-values under the null and alternative hypotheses are, respectively, subscripted by N and A, and h_0^2 and h^2 denote the heritabilities under the null and alternative hypotheses, respectively. The relationship between the heritabilities and the eigenvalues in relation to the function in the summand is depicted in Fig. 1.2. Because of the negative sign outside the summation, eigenvalues less than 1 contribute positively to the ELRT while those greater than 1 decrease the ELRT. This makes intuitive sense since eigenvalues below 1 are direct indications of correlation among individuals. This remarkably simple formula provides the expected test statistic for heritability in pedigrees of arbitrary size and complexity as a function of the easily obtained eigenvalues of the GRK which will often be the pedigree-derived coefficient of relationship matrix. This is the first general formula for pedigree-based heritability testing that we know of. It proves to be a very simple foundation for calculating power to detect heritability.

3.2. Association in the presence of residual heritability

Given that association testing of specific sequence variants is often the focus of genetic analysis of human disease-related phenotypic variation, the sim-plification provided by Equation (1.6) can also be used to derive the ELRT required for fixed-effect testing of marker association. The effect of a sequence variant through a fixed MG effect influencing the mean can be revisualized as a component of genetic variance. Basically, inference on a fixed-effect parameter can be made by examining the perturbation to the variance due to the presence of the fixed effect. This approach is used uni-versally in standard linear regression analysis. Recall that an F statistic is

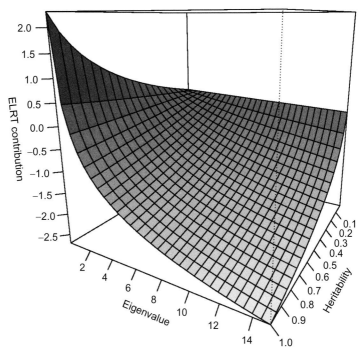

Figure 1.2 Contribution to the ELRT as a function of the eigenvalues and heritabilities. (See Color Insert.)

essentially a ratio of variances. For a standard F-test on a regression coefficient, the variance explained by the regression is compared to the unexplained variance. If the regression parameter is significant (i.e., if the fixed effect is statistically important), it will perturb the variance or, rather, increase the ratio of the explained to the unexplained variance. Similarly, for the ELRT, we can make an inference on the significance of an SNP effect by way of the perturbation to the variance due to the SNP.

The foregoing requires a measure of the VC due to the sequence variant. The modeling of a VC due to a single variant has been addressed by many investigators in human genetics (Blangero et al., 2005; Blangero, Williams, & Almasy, 2000, Boerwinkle et al., 1986; Boerwinkle & Sing, 1986, 1987; Hopper & Matthews, 1982). Here, we will use a simple (but widely biologically valid) model employing additive gene action. Consider now a single diallelic locus representative of a quantitative trait nucleotide (QTN) where the alleles have frequencies p and $(1 - p)$. From classical theory, the QTN variance, denoted by σ_q^2, is known to be

$$\sigma_q^2 = 2p(1-p)a^2, \tag{1.8}$$

where a is the displacement between genotypic means. The QTN-specific heritability, denoted by h_q^2, is therefore given as

$$h_q^2 = \frac{\sigma_q^2}{\sigma_p^2}. \tag{1.9}$$

We use this to define the residual heritability, denoted by h_r^2, which is given as

$$h_r^2 = \frac{h_t^2 - h_q^2}{1 - h_q^2}, \tag{1.10}$$

where h_t^2 is the total heritability. The residual heritability measures the relative amount of additive genetic variation left after accounting for the QTN effect (and any other covariate effects in the model).

Let the covariance matrices under the null and alternative be, respectively, given as

$$\boldsymbol{\Sigma}_N = \left[2\boldsymbol{\Phi}h_t^2 + \left(1-h_t^2\right)\mathbf{I}\right]\sigma_p^2, \tag{1.11}$$

and

$$\boldsymbol{\Sigma}_A = \left[2\boldsymbol{\Phi}h_r^2 + \left(1-h_r^2\right)\mathbf{I}\right]\sigma_p^2\left(1-h_q^2\right). \tag{1.12}$$

Thus, using the eigensimplification of the MVN likelihood, we find for testing association with a sequence variant:

$$
\begin{aligned}
\mathrm{ELRT}\left[h_q^2 > 0 : \mathrm{pedigrees}\right] &= E\left[-2\left(\ln L\left(h_q^2 = 0\right) - \ln L\left(h_q^2 > 0\right)\right)\right] \\
&= n\ln\sigma_p^2 + \sum \ln\left[1 + h_t^2\left(\lambda_{gi} - 1\right)\right] - n\ln\sigma_p^2 \\
&\quad - n\ln\left(1 - h_q^2\right) - \sum \ln\left[1 + h_r^2\left(\lambda_{gi} - 1\right)\right] \\
&= -n\ln\left(1 - h_q^2\right) + \sum \ln\left[1 + h_t^2\left(\lambda_{gi} - 1\right)\right] \\
&\quad - \sum \ln\left[1 + h_r^2\left(\lambda_{gi} - 1\right)\right] \\
&= \mathrm{ELRT}\left[h_q^2 > 0 : \mathrm{unrelateds}\right] \\
&\quad - \left(\mathrm{ELRT}\left[h_t^2 > 0 : \mathrm{pedigrees}\right]\right. \\
&\quad \left. - \mathrm{ELRT}\left[h_r^2 > 0 : \mathrm{pedigrees}\right]\right),
\end{aligned}
\tag{1.13}
$$

where we express the last form of the statistic in terms of the expected statistics for the QTN-specific heritability in a sample of unrelateds, the total heritability in pedigrees, and the residual heritability in pedigrees, respectively. Equation (1.13) provides for the first time a completely general analytical formula for calculating expected association test statistics (and hence power) in arbitrary pedigrees. This formula obviates the need for extensive computer simulation which has been the usual method for obtaining power for association studies on pedigrees. Consistent with conventional wisdom regarding association testing (Visscher, Andrew, & Nyholt, 2008), the last formulation shows the power to detect association in pedigrees will be less than or at most equal to that in a sample of unrelateds. Equation (1.13) should prove of substantial value in the study design of pedigree-based association studies.

4. POWER AND ASYMPTOTIC RELATIVE EFFICIENCY

4.1. Power

Power can be computed as the probability integral from the point on the alternative distribution corresponding to the nominal significance level or alpha (on the null distribution) to the upper limit of the alternative distribution at positive infinity. Since the total probability of any distribution is 1, power can be conveniently computed as

$$
\Pr\left(h^2 > 0\right) = \int_{\chi^2_{\alpha;\nu,\xi=0}}^{\infty} d\chi^2_{\nu,\xi} = \int_0^{\infty} d\chi^2_{\nu,\xi} - \int_0^{\chi^2_{\alpha;\nu,\xi=0}} d\chi^2_{\nu,\xi}
$$

$$
= 1 - \int_0^{\chi^2_{\alpha;\nu,\xi=0}} d\chi^2_{\nu,\xi} = 1 - \beta, \tag{1.14}
$$

where the distribution under the alternative hypothesis is the noncentral chi-square distribution, denoted by $\chi^2_{\nu,\xi}$, ν is the degrees of freedom (d.f.) parameter, ξ is the noncentrality parameter (NCP), $\chi^2_{\alpha;\nu,\xi=0}$ is the point on the noncentral chi-square distribution corresponding to the $100(1-\alpha)$ percentage point on the distribution under the null hypothesis, and $\beta = \int_0^{\chi^2_{\alpha;\nu,\xi=0}} d\chi^2_{\nu,\xi}$ is the probability of making a type 2 error (with apologies for the double use of beta).

When $\xi = 0$, the noncentral chi-square degenerates to the usual chi-square, which is the distribution of the test statistic under the null hypothesis. For standard test cases (e.g., regression coefficients), alpha is the nominal significance level, and so the threshold value for the variate corresponding to the significance level is given as

$$\alpha = 0.05 \le \Pr(\beta = 0) = \Pr\left(\chi_1^2\right),$$

which gives a threshold chi-square of $\chi_1^2 = \chi_{\alpha;v,\xi=0}^2 \cong 3.84146$. This is modified, however, under nonstandard test cases, as in a null hypothesis on the heritability, where the null lies on a boundary of the parameter space. For such cases (and assuming that the variates are independently and identically distributed (i.i.d.)), it is known that the statistic is asymptotically distributed as follows (Chernoff, 1954; DasGupta, 2008; Dominicus, Skrondal, Gjessing, Pedersen, & Palmgren, 2006; Giampaoli & Singer, 2009; Miller, 1977; Self & Liang, 1987; Stram & Lee, 1994; Verbeke & Molenberghs, 2003; Visscher, 2006):

$$\text{LRT} \sim \frac{1}{2}\chi_0^2 + \frac{1}{2}\chi_1^2,$$

which is a 50:50 mixture of a variate with a point mass at 0, denoted by χ_0^2, and a chi-square with 1 d.f., denoted by χ_1^2. Consequently, this upwardly modifies the effective test size:

$$\alpha = 0.05 \le \Pr\left(h^2 = 0\right) = \frac{1}{2}\Pr\left(\chi_0^2\right) + \frac{1}{2}\Pr\left(\chi_1^2\right) \Rightarrow 2\alpha = 1.0 \le 0 + \Pr\left(\chi_1^2\right),$$

which gives a threshold chi-square of $\chi_1^2 = \chi_{2\alpha;v,\xi=0}^2 \cong 2.70554$ (Visscher, 2006). We will revisit this asymptotic distribution of the LRT in a later section of the chapter and show that it is generally conservative.

There are two general methods to calculate power in likelihood analysis owing to the fact that there are two approximations of the NCP for the noncentral chi-square statistic (Brown, Lovato, & Russell, 1999). The older of the two approximations was first derived by Wald (1943) and is equal to the Wald statistic. Although the Wald statistic approximation to the NCP has been commonly used in statistical genetics (Blangero et al., 2001; Williams & Blangero, 1999a,1999b), the requirement of the expected Fisher information matrix makes it burdensome to compute. For the second, work by several investigators has shown that a reasonable NCP approximation is provided by the ELRT (Brown et al., 1999; Liu, 1999; Rijsdijk, Hewitt, & Sham, 2001; Self, Mauritsen, & Ohara, 1992; Sham, Cherny, Purcell, & Hewitt, 2000; Sham, Purcell, Cherny, & Abecasis, 2002). It will be more convenient to use the ELRT in the ensuing power analysis.

4.2. Asymptotic relative efficiency

The concept of asymptotic relative efficiency (ARE) is closely related to power. For two test statistics, denoted by T_1 and T_2, the ARE is defined as the ratio, n_1/n_2, where n_1 and n_2 are the respective theoretical sample sizes for T_1 and T_2 to attain the same power at the same alpha against the same alternative (DasGupta, 2008). Currently, there is no known analytic formula to compute these theoretical sample sizes, but several estimates of the ratio have been developed, one of which will be used here, namely, the Pitman ARE (Pitman, 1948, cited in Noether, 1950, 1955), denoted as e_p. We give the definition of the Pitman ARE for comparing T_1 to T_2 as (DasGupta, 2008):

$$e_p = \left(\lim_{n \to \infty} \frac{\sqrt{n}\sigma_{\theta_1}}{\mu_{\theta_1}} \middle/ \lim_{n \to \infty} \frac{\sqrt{n}\sigma_{\theta_2}}{\mu_{\theta_2}} \right)^2 = \frac{\sigma_{\theta_1}^2}{\sigma_{\theta_2}^2} \left(\frac{\mu_{\theta_2}}{\mu_{\theta_1}} \right)^2 = \frac{\sigma_{\theta_1}^2}{\sigma_{\theta_2}^2} \left(\frac{\hat{\theta}_2}{\hat{\theta}_1} \right)^2, \quad (1.15)$$

where the components are subscripted by test number, and the asymptotic parameter means are equivalent to the parameter MLEs. For many cases, the parameter standard errors (and hence their variances) are test specific, whereas the parameter means or MLEs are asymptotically equivalent (DasGupta, 2008). Thus, for such cases, including the current situation, e_p is given as the ratio of the variances.

The latter most form leads directly to the following useful result:

$$e_p = \frac{\sigma_{\theta_1}^2}{\sigma_{\theta_2}^2} \left(\frac{\hat{\theta}_2}{\hat{\theta}_1} \right)^2 = \frac{\hat{\theta}_2^2}{\sigma_{\theta_2}^2} \middle/ \frac{\hat{\theta}_1^2}{\sigma_{\theta_1}^2} = \frac{W_2}{W_1} = \frac{\mathrm{NCP}_2}{\mathrm{NCP}_1}, \quad (1.16)$$

where W is the Wald statistic. We emphasize that the direction of the comparison is still T_1 to T_2 despite the fact that the direction in terms of NCPs is NCP_2 to NCP_1. This formulation of the ARE has been commonly used in human statistical genetics to compare the relative power of two different tests (Bhattacharjee et al., 2010; Kim, Gordon, Sebat, Ye, & Finch, 2008; Visscher et al., 2008; Visscher & Duffy, 2007; Yang, Wray, & Visscher, 2010).

4.2.1 Heritability

Equation (1.16) suggests we can use the ELRT in a simple alternative measure of the Pitman ARE since it measures the NCP:

$$e_p = \frac{\mathrm{ELRT}_2}{\mathrm{ELRT}_1} = \frac{\sum \ln\left[1 + h_2^2 \left(\lambda_{\mathrm{gi2}} - 1 \right) \right]}{\sum \ln\left[1 + h_1^2 \left(\lambda_{\mathrm{gi1}} - 1 \right) \right]}, \quad (1.17)$$

where the statistics, heritabilities, and eigenvalues are subscripted by test number.

4.2.2 Association

Using the ELRT for association, we have the following alternative:

$$e_p = \frac{\ln\left(1 - h_{q2}^2\right) + \sum \ln\left[1 + h_{t2}^2\left(\lambda_{gi2} - 1\right)\right] - \sum \ln\left[1 + h_{r2}^2\left(\lambda_{gi2} - 1\right)\right]}{\ln\left(1 - h_{q1}^2\right) + \sum \ln\left[1 + h_{t1}^2\left(\lambda_{gi1} - 1\right)\right] - \sum \ln\left[1 + h_{r1}^2\left(\lambda_{gi1} - 1\right)\right]}.$$

(1.18)

Equations (1.17) and (1.18) provide simple formula for directly comparing different pedigree designs for optimality of inference.

5. UTILITY OF EIGENSIMPLIFICATION FOR THE POLYGENIC MODEL

5.1. Analytic eigenvalues for pedigree-derived GRKs

Our analytical results clearly demonstrate the primacy of the eigenvalues distribution for a given GRK as the focal determinant of power to detect heritability. For canonical relationships and simple pedigree structures such as those shown in Fig. 1.3, the eigenvalues of the pedigree-derived GRK can be analytically determined. Such analytical determinations are extremely useful when considering theoretical issues of study design or when trying to determine what type of family would be best for recruitment in a given proposed study. Table 1.1 shows the analytical eigenvalues for those common pedigree structures depicted in Fig. 1.3. For more complex extended families such as the one depicted in Fig. 1.4 (this is an actual family from our San Antonio Family Heart Study (SAFHS) sample that has undergone WGS), the eigenvalues must be numerically determined by spectral decomposition of the pedigree-derived GRK. Figure 1.5 shows a histogram of the eigenvalues of the relationship matrix for this large pedigree that were obtained numerically. Recall that eigenvalues less than 1 contribute positively to the test statistic for heritability. As can be seen for relationship structures with more than two individuals in Table 1.1 and Fig. 1.5, eigenvalues less than 1 are always more frequent. A slight problem arises for the case of monozygotic (MZ) twins in that the ELRT (and, in fact, the MVN likelihood function) becomes degenerate at heritability exactly equal to 1. This problem can be dealt with by bounding the heritability slightly less than 1.

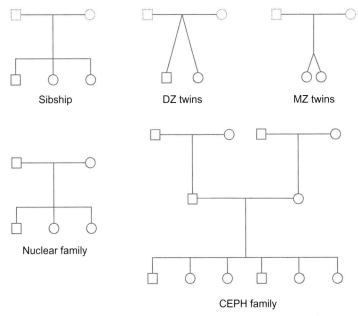

Sibship   DZ twins   MZ twins

Nuclear family

CEPH family

Figure 1.3 Some simple relationship and pedigree structures. (See Color Insert.)

5.2. Power functions for heritability and association

To evaluate the accuracy of the theory, we analyzed parametric bootstrap simulations of a quantitative trait sampled from the SAFS example pedigree (EP). Basically, using the simulation modules in our SOLAR software (Almasy & Blangero, 1998), we simulated heritabilities across the parameter space and examined our empirical power to obtain significant evidence for genetic factors. Using 10,000 simulations including 10 copies of the SAFHS EP in each simulation, we obtained a close correspondence between theory and empirical observation. Figure 1.6 demonstrates how close the theoretical ELRTs come to a 6th order polynomial fit of the simulated LRTs. Clearly, our very simple formula for the ELRT is suitably accurate for general use. Table 1.1 also shows the ELRT per relationship unit and per individual for two levels of heritability (0.30 and 0.70). Our results show a dependence of the ELRT upon the total heritability and the pedigree eigenvalues. Similarly, in Fig. 1.7, we plot on the left panel the ELRTs for four of the different relationship structures, namely, MZ twins (noted as MZ), nuclear families with three siblings (NF), CEPH-style families with six siblings (CEPH), and the SAFHS EP, all scaled to 250 individuals for easy comparison. In the right panel of Fig. 1.7, we show the scaled power functions for heritability estimation for the same pedigree

Table 1.1 Analytic eigenvalues for various relationship structures

Relationship structure	Additive genetic eigenvalues	ELRT per relationship unit	
		$h^2 = 0.3$	$h^2 = 0.7$
MZ twins	2 0	$0.09\ (0.05)^{\text{a}}$	$0.67\ (0.34)$
Sib pair	$3/2$ $1/2$	$0.02\ (0.005)$	$0.13\ (0.03)$
Sibship	$(n_s + 1)/2$ $(n_s - 1)\{1/2\}^{\text{b}}$	$0.18\ (0.04)^{\text{c}}$	$0.85\ (0.17)$
Relative pair in GRK, $\mathbf{K} = \{K_{ij}\}$	$1 + K_{ij}$ $1 - K_{ij}$	$0.006\ (0.003)^{\text{d}}$	$0.03\ (0.015)$
Nuclear family	$(n_s + 3)/4 \pm \sqrt{2n_s + (n_s - 1)^2/4}/2$ 1 $(n_s - 1)\{1/2\}$	$0.17\ (0.03)$	$0.90\ (0.18)$
CEPH family	$(2)\{1\}$ $(n_s)\{1/2\}$ $1 \pm \sqrt{2}/2$ $(n_s + 4)/4 \pm \sqrt{2(n_s + 1) + n_s^2/4}/2$	$0.56\ (0.05)$	$2.54\ (0.21)$
Extended pedigree	Eigenvalues of $\mathbf{K}$	$10.30^{\text{e}}\ (0.06)$	$40.62\ (0.24)$

[a]The number in parentheses is the scaled individual contribution to the ELRT.
[b]We use the symbology $(x)\{y\}$ to denote x units of value y. Otherwise, the operators are to be interpreted in the usual manner. For example, the first sibship entry means there is one eigenvalue at that value, and the second entry means that there are $(n_s - 1)$ eigenvalues (n_s being the number of sibs), each one equal to $1/2$.
[c]For 5 sibs.
[d]For grandparent–grandchild, or avuncular, or half-sib relationships.
[e]For the extended pedigree in Fig. 1.4 ($N = 171$).

structures. Notably, the extended family design is most powerful in the region of the null. However, for heritabilities above approximately 0.47, the MZ design becomes most powerful. Our results indicate that the conventionally and widely held belief that MZ twins constitute the most powerful design for estimating heritability is not true in the most important part of the parameter space (i.e., in the local area of the null hypothesis).

We examined similar study design comparisons in relation to association testing using Equation (1.13). For a total heritability of 0.1, we plotted the association ELRTs for the four relationship structures and power to detect

San Antonio Family Study: Pedigree 1

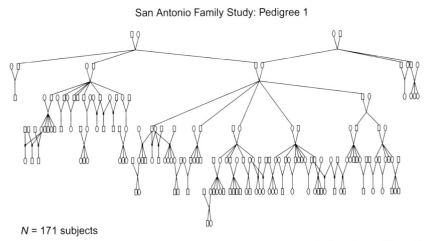

N = 171 subjects

Figure 1.4 A San Antonio Family Heart Study (SAFHS) extended pedigree (*N* = 171 individuals).

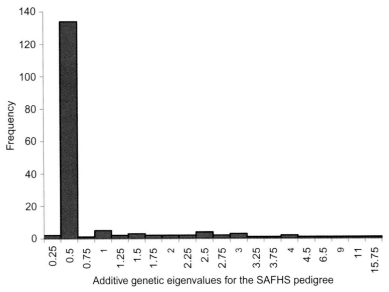

Additive genetic eigenvalues for the SAFHS pedigree

Figure 1.5 Numerically estimated eigenvalues of the SAFHS extended pedigree. (See Color Insert.)

association for the same fixed sample sizes in Fig. 1.8. Figure 1.8 shows that, for this fixed low total heritability and a reasonable range of QTN-specific heritabilities, power to detect associations is greatest in unrelated samples as expected. Loss of power is greatest in the extended pedigree due to the

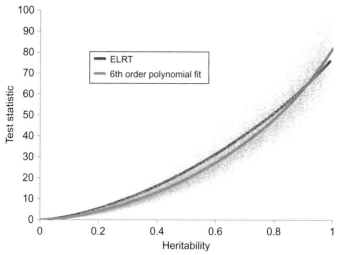

Figure 1.6 Parametric bootstrap of the LRT distribution. Simulation of 19,000 LRTs where the generating model is for heritability estimation using the SAFHS extended pedigree. 6th order polynomial fit (blue line). ELRT computed for the SAFHS (red line). (See Color Insert.)

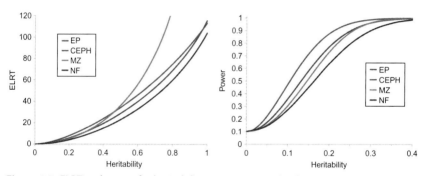

Figure 1.7 ELRT and power for heritability estimation. For both panels: SAFHS extended pedigree (EP) (blue line), CEPH pedigree (red line), monozygotic twins (MZ) (green line), and nuclear family (NF) (purple line) all scaled to 250 individuals. (See Color Insert.)

substantial correlation between subjects; however, even for this design, the loss of power is low. Figure 1.9 shows the effect of total heritability on power to detect association for the extended pedigree. Power to detect association is influenced by total heritability with the power loss being maximized at a residual heritability of 0.50. Interestingly, power loss as seen in the ELRT is

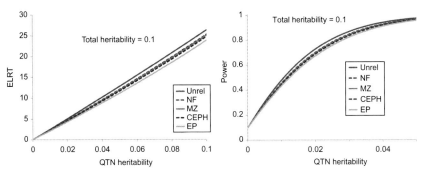

Figure 1.8 ELRT and power for association testing. For both panels: Unrelateds (blue line), nuclear family (NF) (red dashed line), monozygotic twins (MZ), (green line), CEPH pedigree (purple dashed line), and SAFHS extended pedigree (EP) (yellow). (See Color Insert.)

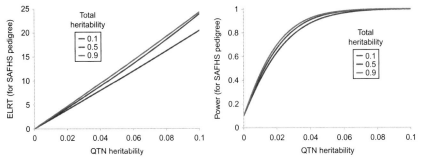

Figure 1.9 Effect of varying the total heritability on ELRT and power for association testing all scaled to 250 individuals. Total heritabilities equal to: 0.1 (blue line), 0.5 (red line), and 0.9 (green line). (See Color Insert.)

minimal both near the null region for heritability (as expected) and somewhat counterintuitively near the maximum of heritability (at 1). Regardless, our theoretical results show little loss of power in the association analysis of even large and complex pedigrees. Furthermore, when considering the increased focus on the analysis of rare variants, power is actually substantially increased in large pedigrees due to the accumulation of multiple copies (and hence, increased genotypic variance) of rare variants incurred through Mendelian transmission in variant-harboring lineages.

5.3. Asymptotic relative efficiency

We also calculated the AREs for comparing pedigree-based designs for association analysis. Figure 1.10 shows all ARE comparisons relative to unrelateds in Fig. 1.10, again scaled to 250 individuals. Our results are

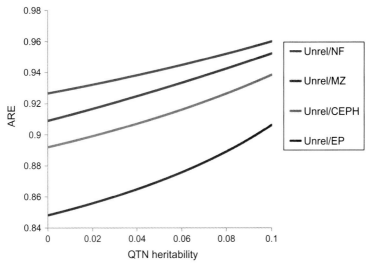

Figure 1.10 Pitman Asymptotic Relative Efficiency (ARE) for unrelated in relation to various family structures. Unrelateds in comparison to: nuclear family (NF) (blue line), monozygotic twins (MZ) (red line), CEPH family (green), and SAFHS extended pedigree (EP) (purple line). (See Color Insert.)

consistent with those from a study by Visscher et al. (2008). They found that there is, in fact, little power loss on comparing the power to detect association in a sample of unrelateds versus in a sample of relatives. In fact, the power loss becomes even smaller at higher total heritabilities (analyses not shown) as suggested also in Fig. 1.9.

5.4. Inadequacy of the asymptotic LRT distribution for VC testing

As we briefly discussed earlier, the asymptotic distribution of the LRT for testing the null hypothesis with regard to a VC is given by a 50:50 mixture of χ_0^2, which denotes a chi-square random variable with a point mass at 0, and of χ_1^2, a chi-square with 1 d.f. However, this is the appropriate distribution only if the data are i.i.d. (Crainiceanu, 2008; Crainiceanu & Ruppert, 2004a, 2004b, 2004c; Crainiceanu, Ruppert, Claeskens, & Wand, 2005; Crainiceanu, Ruppert, & Vogelsang, 2003). Unfortunately, for most VC models in use in pedigree analyses including the ones under discussion here, this assumption is violated, and this departure has been shown to generate skewed mixture distributions that have an increased frequency of χ_0^2 (i.e., an increased incidence of test statistics of zero). Researchers have found that the true frequency of χ_0^2 can range from 0.65 to as high as 0.96(!)

(Crainiceanu & Ruppert, 2004a; Crainiceanu et al., 2003; Kuo, 1999; Pinheiro & Bates, 2000; Shephard, 1993; Shephard & Harvey, 1990). This means that the traditional theory for nonstandard cases can be severely conservative and hence show a loss of power. Consequently, Crainiceanu and colleagues developed an elegant and useful theory to recover the appropriate distribution (Crainiceanu, 2008; Crainiceanu & Ruppert, 2004a, 2004b, 2004c; Crainiceanu et al., 2003, 2005). In fact, like us, they also employ a spectral representation of the likelihood function to obtain a simplified LRT. Tailoring their theory to the present situation, let π_i and λ_{gi} be the eigenvalues of $\mathbf{K}^{1/2}\mathbf{P}\mathbf{K}^{1/2}$ and $\mathbf{K}$, respectively, where $\mathbf{P} = \mathbf{I} - \mathbf{X}(\mathbf{X}'\mathbf{X})^{-1}\mathbf{X}'$. For the heritability problem and large n, these two matrices (and their eigenvalues) tend toward equality. Then, if the true value of the heritability is given by h_0^2, their expression for the LRT is given as

$$\text{LRT} = \sup_{h^2}\left[n\ln\left\{1 + \frac{N}{D}\right\} - \sum\ln\left(\frac{1 + \frac{h^2\lambda_{gi}}{e^2}}{1 + \frac{h_0^2\lambda_{gi}}{e_0^2}}\right)\right], \quad (1.19)$$

where $e^2 = 1 - h^2$, $e_0^2 = 1 - h_0^2$, $N = \sum(\pi_i/\psi)(h^2 e_0^2 - h_0^2 e^2)\omega_i^2$, $D = \sum(e^2/\psi)(e_0^2 + h_0^2\pi_i)\omega_i^2 + \sum_{i=1}^{n-j}\omega_i^2$, $\psi = (e^2 + h^2\pi_i)e_0^2$, the maximization is with regard to h^2, the summations are over all n values unless explicitly noted, and the ω_is are independent standard normal random variables, $N(0,1)$, similar to the $\boldsymbol{\tau}$ vector in Equation (1.5). The probability of χ_0^2 is equal to the probability that Equation (1.19) has a global maximum at $h^2 = 0$. To approximate this probability, Crainiceanu and colleagues recommended computing the probability of a local maximum at $h^2 = 0$ for a given sample size. This probability is given as

$$P\left\{\frac{\sum\pi_i\omega_i^2}{\sum_{i=1}^{n-j}\omega_i^2}\right\} \le \frac{1}{n}\sum\lambda_{gi} \Rightarrow P\left\{\frac{\sum\pi_i\omega_i^2}{\sum_{i=1}^{n-j}\omega_i^2}\right\} \le 1 \quad (1.20)$$

Unfortunately, it appears that the true distribution of the LRT in finite situations is determined by the asymptotic distribution of the eigenvalues of the matrices involved. Thus, for every study and every covariate configuration, a separate LRT distribution should be examined. While this sounds formidable, these formulae suggest a simple and very rapid method for obtaining the true expected LRT distribution using simulation. We used the R program RLRsim (Scheipl & Bolker, 2012; Scheipl, Greven, & Küchenhoff, 2008) to simulate the LRTs using the spectral representation of the LRT in Equation (1.19) for our extended SAFHS pedigree

(Fig. 1.4). Fitting a mixture of a binomial and a χ_1^2 distribution as suggested in Crainiceanu (2008) and Greven, Crainiceanu, Küchenhoff, and Peters (2008), we estimated a true mixing proportion of 0.57:0.43 and a multiplicative correction to the χ_1^2 distribution of approximately 0.905. The true cut-off is closer to 2.3 than the asymptotic theory prediction of 2.7. Thus, as expected, reliance on the asymptotic theory for nonstandard test cases will be conservative. Experimentation shows that there is much less of an effect on association inference and that the nonstandard asymptotic theory holds well.

6. ANALYSIS OF EMPIRICAL GRKs

Empirical GRKs have proved to be quite useful in the development of novel statistical genetic methods. Visscher and colleagues (Yang, Manolio, et al., 2011) have used empirical GRKs to even extract quantitative genetic information from "unrelated" individuals by exploiting deep ancestries. For example, one way to greatly reduce the problem of multiple hypothesis testing when analyzing a prohibitively large number of SNPs is to estimate **K** from the set of SNPs and to use it to model a VC reflective of the aggregate effects of the SNPs (Wu et al., 2011). Obviously, the multiple testing problem is amplified in the setting of WGS data analysis, which accordingly increases the utility of a method that produces single degree of freedom tests. This approach has been applied to the analysis of several complex traits (Yang, Benyamin, et al., 2010; Yang, Lee, et al., 2011; Yang, Manolio, et al., 2011; Yang, Wray, et al., 2010;), including height, body mass index, von Willebrand factor, and QT interval, and to schizophrenia (Lee et al., 2012). This idea could be extended to computing heritabilities on a chromosomal or a gene segment basis (Yang, Lee, et al., 2011; Yang, Manolio, et al., 2011). One could also leverage this approach to compute the **K** relevant for a metabolic pathway and to estimate pathway-specific VCs (Almeida et al., 2012). We have shown that it is possible to accurately recover both total and local (i.e., QTL-specific) heritability estimates by using only empirical GRKs in a known pedigree situation (Day-Williams, Blangero, Dyer, Lange, & Sobel, 2011). However, see Hill and Weir (2011) for cautionary caveats on the potential loss of accuracy in the empirical relatedness of remote relatives.

Empirical GRKs have the potential to make significant contributions to the statistical genetic analysis of complex traits and diseases. As an example of their use and to illustrate potential problems, we estimated a GRK using the

GCTA software (Yang, Lee et al., 2011). Asymptotically, this procedure should yield a test of heritability that is consistent with that of the underlying average coefficient of relationship matrix. Again, we focused on the extended pedigrees of the general complexity as that shown in Fig. 1.4. We employed the WGS single nucleotide variant frequency spectrum information (for half the genome, specifically the odd-numbered autosomes) available on 20 SAFS pedigrees including 852 individuals with data utilized in the most recent Genetic Analysis Workshop 18 (Almasy et al., in press) that this pedigree was part of, and simulated 4.1M SNVs with minor allele frequencies > 0.01 for our GRK estimates. The resulting kernel is positive semi-definite (PSD). However, Fig. 1.11 shows that the critical eigenvalue distribution is different from that observed for the true pedigree-derived relationship matrix. Specifically, the GCTA leading eigenvalues are deflated which occurs when overall correlation amongst individuals is underestimated. Based on our Equation 1.7, this should lead to inflation of the test statistic. We performed a simulation experiment to test the influence of having an empirical GRK for heritability estimation. Using our SOLAR software, we obtained one million replicates under the null hypothesis of no heritability using the expected kernel given by the observed SAFHS pedigree structure. We then analyzed the simulated quantitative traits under the true generating model (using the pedigree-derived GRK) and under a model using the empirical GRK. We used the approach described in Equations (1.19) and (1.20) to obtain the true null distribution under the generating model. Using this cut-off, we observed that type 1 error for the GCTA GRK is inflated with a false positive rate of 0.053, a 6% increase in error. The type 1 error worsens to a 10% excess for a more stringent significance cut-off of 0.001.

Whilst the GCTA GRK was PSD, many empirical GRK estimation procedures can lead to non-PSD kernels. Non-PSD matrices have been a bane in statistics in general and statistical genetics in particular for decades now. It was previously observed that a non-PSD covariance matrix can substantially bias heritability estimates (Hayes & Hill, 1980, 1981; Hill & Thompson, 1978). For non-PSD matrices, the larger eigenvalues are biased upward and the smaller eigenvalues are biased downward (Hayes & Hill, 1980, 1981; Hill & Thompson, 1978; Meyer & Kirkpatrick, 2008). Our *ELRT* formula (Equation 1.7) for heritability shows that negative eigenvalues will inflate the test statistic and hence may lead to increased type 1 error under the null hypothesis of no heritability. Because of the shape of the *ln* function, the negative eigenvalues have a disproportionate effect on

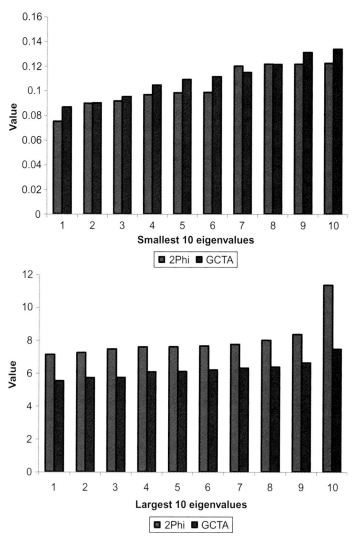

Figure 1.11 Comparison of the smallest and largest eigenvalues computed from the true pedigree-derived (2Phi) and GCTA relationship kernels. For both panels: 2Phi (blue columns), GCTA (red columns). (See Color Insert.)

the total test statistic. Our results suggest that it is important to constrain empirical GRKs to be PSD. This will typically require some type of post-processing of an estimated GRK. For example, an early approach was to correct such non–PSD matrices by adjusting the eigenvalue distribution so that the smallest eigenvalue equals 0 (Hayes & Hill, 1980, 1981; Kirkpatrick & Lofsvold, 1992). Subsequent approaches instead aimed to obtain a better estimate of the covariance matrix while guaranteeing the matrix to be PSD, and are reviewed in Meyer and Kirkpatrick (2008, 2010) and Meyer (2011). Our theoretical results suggest that care should be given when choosing empirical GRKs.

7. DISCUSSION

The main theme of this work has been on the practical utility and unifying value of our eigensimplification of the polygenic VC likelihood function. The eigensimplification approach enables much more rapid computations that are equivalent to those required in samples of unrelateds after initial transformation, which is a highly practical benefit in this new era of high-dimensional NGS data analysis. Importantly, it also led to elegant theoretical advances in regard to the ELRT, power analysis, and the analysis of GRKs. Our general formulae related to power to detect heritability in arbitrary pedigrees represent a solution to a difficult problem that has typically been handled using computer simulation. Our formulae unequivocally show that the critical parameters for power to detect heritability involve the eigenvalues of the pedigree-relationship matrix. We have also used our approach to examine the expected power of arbitrary pedigrees for detecting associations. Again, to assess power for association testing in pedigrees, investigators have typically been required to rely on cumbersome simulation strategies. Our formulae now allow rapid analytical evaluation of different study designs. Our simple formulation of the ELRT for association in pedigrees also shows exactly how power is lost due to non-independence between relatives. We also show that this power loss is relatively minor even for our largest most complex pedigree analyzed. Given that we are now entering an era where association studies of rare variants in large pedigrees is likely to rapidly increase, our results will be useful for aiding rational study design in the genetic dissection of complex phenotypes.

Our analytical approach is not without its own share of weaknesses. One major criticism that may turn out to be an inroad for future advance is that the exact spectral decomposition approach is limited to VC models with only two VCs. This is because only two kernels at a time can be simultaneously diagonalized, whereas generalization to an arbitrary number of matrices requires numerical approximation (Flury & Gautschi, 1986). Thus, neither is the approach immediately able to incorporate a linkage VC (certainly not analytically), which is known to greatly improve upon the overall VC statistical genetic model, nor is it possible to extend the approach to model genotype-by-environment interaction using an additional VC. We are now working with empirical eigensimplification approaches which substantially reduce computation but not to the extent of that observed under the simple polygenic VC model, nor do they lead to such obvious insights into the canonical determinants of power. Notwithstanding these important criticisms, we believe that the eigensimplification approach of the classical additive polygenic model will lead to important empirical and possibly even additional theoretical discoveries.

Finally, all of the procedures discussed in this work have been (or will be in the near future) implemented into our general statistical genetic software package, SOLAR, available from http://txbiomed.org/departments/genetics/genetics-detail?r=37.

ACKNOWLEDGMENTS

The development of the analytical methods and software used in this study was supported by NIH grant R37 MH059490. Data collection for the San Antonio Family Heart Study was supported by NIH grant R01 HL045522. We are grateful to the participants of the San Antonio Family Heart Study for their continued involvement. The GAW18 data are funded by NIH grant R01 GM031575 and the WGS data used in GAW18 were funded by NIH grants U01 DK085524, U01 DK085584, U01 DK085501, U01 DK085526, and U01 DK085545. The AT&T Genomics Computing Center supercomputing facilities used for this work were supported in part by a gift from the AT&T Foundation and with support from the National Center for Research Resources Grant Number S10 RR029392.

REFERENCES

1000 Genomes Project Consortium, Abecasis, G. R., Auton, A., Brooks, L. D., DePristo, M. A., Durbin, R. M., et al. (2012). An integrated map of genetic variation from 1,092 human genomes. *Nature, 491*, 56–65.

Abadir, K. M., & Magnus, J. R. (2005). *Matrix algebra*. Cambridge: Cambridge University Press.

Almasy, L., & Blangero, J. (1998). Multipoint quantitative-trait linkage analysis in general pedigrees. *American Journal of Human Genetics, 62*, 1198–1211.

Almasy, L., & Blangero, J. (2010). Variance component methods for analysis of complex phenotypes. *Cold Spring Harbor Protocols, 2010*(5) pdb.top77.

Almasy, L., Dyer, T. D., Peralta, J. M., Jun, G., Fuchsberger, C., Almeida, M. A., et al. (in press). Data for genetic analysis workshop 18: Human whole genome sequence, blood pressure, and simulated phenotypes in extended pedigrees. *Genetic Epidemiology*.

Almeida, M., Peralta, J., Farook, V., Puppala, S., Duggirala, R., & Blangero, J. (2012). Random effect burden tests to screen gene pathways. In: *Genetic Analysis Workshop 18, Stevenson, WA, October 13–17*.

Anderson, T. W. (1984). *An introduction to multivariate statistical analysis* (2nd ed.). New York: John Wiley & Sons.

Anderson, T. W., & Olkin, I. (1985). Maximum-likelihood estimation of the parameters of a multivariate normal distribution. *Linear Algebra and its Applications, 70*, 147–171.

Bhattacharjee, S., Wang, Z., Ciampa, J., Kraft, P., Chanock, S., Yu, K., et al. (2010). Using principal components of genetic variation for robust and powerful detection of gene-gene interactions in case-control and case-only studies. *American Journal of Human Genetics*, *86*, 331–342.

Blangero, J. (2004). Localization and identification of human quantitative trait loci: King Harvest has surely come. *Current Opinion in Genetics & Development*, *14*, 233–240.

Blangero, J., Göring, H. H. H., Kent, J. W., Jr., Williams, J. T., Peterson, C. P., Almasy, L., et al. (2005). Quantitative trait nucleotide analysis using Bayesian model selection. *Human Biology*, *77*, 541–559.

Blangero, J., Williams, J. T., & Almasy, L. (2000). Quantitative trait locus mapping using human pedigrees. *Human Biology*, *72*, 35–62.

Blangero, J., Williams, J. T., & Almasy, L. (2001). Variance component methods for detecting complex trait loci. In D. C. Rao & M. A. Province (Eds.), *Genetic dissection of complex traits. Advances in genetics*, Vol. 42, (pp. 151–181). New York: Academic Press.

Boerwinkle, E., Chakraborty, R., & Sing, C. F. (1986). The use of MG information in the analysis of quantitative phenotypes in man. I. Models and analytical methods. *Annals of Human Genetics*, *50*, 181–194.

Boerwinkle, E., & Sing, C. F. (1986). Bias of the contribution of single-locus effects to the variance of a quantitative trait. *American Journal of Human Genetics*, *39*, 137–144.

Boerwinkle, E., & Sing, C. F. (1987). The use of MG information in the analysis of quantitative phenotypes in man. III. Simultaneous estimation of the frequencies and effects of the apolipoprotein E polymorphism and residual polygenetic effects on cholesterol, betalipoprotein and triglyceride levels. *Annals of Human Genetics*, *51*, 211–226.

Brown, B. W., Lovato, J., & Russell, K. (1999). Asymptotic power calculations: Description, examples, computer code. *Statistics in Medicine*, *18*(22), 3137–3151.

Chernoff, H. (1954). On the distribution of the likelihood ratio. *Annals of Mathematical Statistics*, *25*, 573–578.

Crainiceanu, C. M. (2008). Likelihood ratio testing for zero variance components in linear mixed models. In D. B. Dunson (Ed.), *Random effect and latent variable model selection* (pp. 3–17). New York: Springer.

Crainiceanu, C. M., & Ruppert, D. (2004a). Restricted likelihood ratio tests in nonparametric longitudinal models. *Statistica Sinica*, *14*, 713–729.

Crainiceanu, C. M., & Ruppert, D. (2004b). Likelihood ratio tests in linear mixed models with one variance component. *Journal of Royal Statistical Society, Series B*, *66*, 165–185.

Crainiceanu, C. M., & Ruppert, D. (2004c). Likelihood ratio tests for goodness-of-fit of a nonlinear regression model. *Journal of Multivariate Analysis*, *91*, 35–52.

Crainiceanu, C. M., Ruppert, D., Claeskens, G., & Wand, M. P. (2005). Exact likelihood ratio tests for penalized splines. *Biometrika*, *92*, 91–103.

Crainiceanu, C. M., Ruppert, D., & Vogelsang, T. J. (2003). Some properties of the likelihood ratio tests in linear mixed models. Available at http://legacy.orie.cornell.edu/davidr/papers/zeroprob_rev01.pdf.

DasGupta, A. (2008). *Asymptotic theory of statistics and probability*. New York: Springer.

Day-Williams, A. G., Blangero, J., Dyer, T. D., Lange, K., & Sobel, E. M. (2011). Linkage analysis without defined pedigrees. *Genetic Epidemiology*, *35*, 360–370.

Dempster, A. P., Patel, C. M., Selwyn, M. R., & Roth, A. J. (1985). Statistical and computational aspects of mixed model analysis. *Applied Statistics*, *33*, 203–214.

Dominicus, A., Skrondal, A., Gjessing, H. K., Pedersen, N. L., & Palmgren, J. (2006). Likelihood ratio tests in behavioral genetics: Problems and solutions. *Behavior Genetics*, *36*, 331–340.

Dyer, T. D., Diego, V. P., Kent, J. W., Jr., Göring, H. H. H., & Blanger, J. (2009). Rapid exact likelihood-based quantitative trait association analysis in large pedigrees. In: *American Society of Human Genetics Annual Meeting, Honolulu, HI, October 20–24*.

Fisher, R. A. (1918). The correlation between relatives on the supposition of Mendelian inheritance. *Transactions of the Royal Society of Edinburgh, 52,* 399–433.

Flury, B. N., & Gautschi, W. (1986). An algorithm for simultaneous orthogonal transformation of several positive definite symmetric matrices to nearly diagonal form. *SIAM Journal on Scientific and Statistical Computing, 7,* 167–184.

Giampaoli, V., & Singer, J. M. (2009). Likelihood ratio tests for variance components in linear mixed models. *Journal of Statistical Planning and Inference, 139,* 1435–1448.

Greven, S., Crainiceanu, C. M., Küchenhoff, H., & Peters, A. (2008). Restricted likelihood ratio testing for zero variance components in linear mixed models. *Journal of Computational and Graphical Statistics, 17,* 870–891.

Hayes, J. F., & Hill, W. G. (1980). A reparameterization of a genetic selection index to locate its sampling properties. *Biometrics, 36,* 237–248.

Hayes, J. F., & Hill, W. G. (1981). Modification of estimates of parameters in the construction of genetic selection indices ('bending'). *Biometrics, 37,* 483–493.

Hill, W. G., & Thompson, R. (1978). Probabilities of non-positive definite between-group or genetic covariance matrices. *Biometrics, 34,* 429–439.

Hill, W. G., & Weir, B. S. (2011). Variation in actual relationship as a consequence of Mendelian sampling and linkage. *Genetics Research, 93,* 47–64.

Hopper, J. L., & Matthews, J. D. (1982). Extensions to multivariate normal models for pedigree analysis. *Annals of Human Genetics, 46,* 373–383.

Kang, H. M., Sul, J. H., Service, S. K., Zaitlen, N. A., Kong, S. Y., Freimer, N. B., et al. (2010). Variance component model to account for sample structure in genome-wide association studies. *Nature Genetics, 42,* 348–354.

Kang, H. M., Zaitlen, N. A., Wade, C. M., Kirby, A., Heckerman, D., Daly, M. J., et al. (2008). Efficient control of population structure in model organism association mapping. *Genetics, 178,* 1709–1723.

Kim, W., Gordon, D., Sebat, J., Ye, K. Q., & Finch, S. J. (2008). Computing power and sample size for case-control association studies with copy number polymorphism: Application of mixture-based likelihood ratio test. *PLoS One, 3,* e3475.

Kirkpatrick, M., & Lofsvold, D. (1992). Measuring selection and constraint in the evolution of growth. *Evolution, 46,* 954–971.

Kuo, B.-S. (1999). Asymptotics of ML estimator for regression models with a stochastic trend component. *Econometric Theory, 15,* 24–49.

Lange, K. (2002). *Mathematical and statistical methods for genetic analysis* (2nd ed.). New York: Springer-Verlag.

Lee, S. H., DeCandia, T. R., Ripke, S., Yang, J., Schizophrenia Psychiatric Genome-Wide Association Study Consortium (PGC-SCZ), , International Schizophrenia Consortium (ISC), , et al. (2012). Estimating the proportion of variation in susceptibility to schizophrenia captured by common SNPs. *Nature Genetics, 44,* 247–250.

Liu, B. (1999). *Statistical genomics: Linkage, mapping, and QTL analysis.* Boca Raton: CRC Press.

Meyer, K. (2011). Performance of penalized maximum likelihood in estimation of genetic covariances matrices. *Genetics, Selection, Evolution, 43,* 39.

Meyer, K., & Kirkpatrick, M. (2008). Perils of parsimony: Properties of reduced-rank estimates of genetic covariance matrices. *Genetics, 180,* 1153–1166.

Meyer, K., & Kirkpatrick, M. (2010). Better estimates of genetic covariance matrices by "bending" using penalized maximum likelihood. *Genetics, 185,* 1097–1110.

Miller, J. J. (1977). Asymptotic properties of maximum likelihood estimates in the mixed model of the analysis of variance. *The Annals of Statistics, 5,* 746–762.

Moll, P. P., Powsner, R., & Sing, C. F. (1979). Analysis of genetics and environmental sources of variation in serum cholesterol in Tecumseh, Michigan. V. Variance components estimated from pedigrees. *Annals of Human Genetics, 42,* 343–354.

Noether, G. E. (1950). Asymptotic properties of the Wald-Wolfowitz test of randomness. *Annals of Mathematical Statistics, 21,* 231–246.

Noether, G. E. (1955). On a theorem of Pitman. *Annals of Mathematical Statistics*, *26*, 64–68.

Ott, J., Kamatani, Y., & Lathrop, M. (2011). Family-based designs for genome-wide association studies. *Nature Reviews. Genetics*, *12*, 465–474.

Patterson, H. D., & Thompson, R. (1971). Recovery of inter-block information when block sizes are unequal. *Biometrika*, *58*, 545–554.

Pettofrezzo, A. J. (1978). *Matrices and transformations*. New York: Dover Publications.

Pinheiro, J. C., & Bates, D. M. (2000). *Mixed-effects models in S and S-plus*. New York: Springer.

Rijsdijk, F. V., Hewitt, J. K., & Sham, P. C. (2001). Analytic power calculation for QTL linkage analysis of small pedigrees. *European Journal of Human Genetics*, *9*, 335–340.

Scheipl, F., & Bolker, B. (2012). Package 'RLRsim'. Available at: http://cran.r-project.org/web/packages/RLRsim/index.html.

Scheipl, F., Greven, S., & Küchenhoff, H. (2008). Size and power of tests for zero random effect variance or polynomial regression in additive and linear mixed models. *Computational Statictics and Data Analysis*, *52*, 3283–3299.

Self, S. G., & Liang, K.-Y. (1987). Asymptotic properties of maximum likelihood estimators and likelihood ratio tests under nonstandard conditions. *Journal of the American Statistical Association*, *82*, 605–610.

Self, S. G., Mauritsen, R. H., & Ohara, J. (1992). Power calculations for likelihood ratio tests in generalized linear models. *Biometrics*, *48*, 31–39.

Sham, P. C., Cherny, S. S., Purcell, S., & Hewitt, J. K. (2000). Power of linkage versus association analysis of quantitative traits, by use of variance-components models, for sibship data. *American Journal of Human Genetics*, *66*, 1616–1630 Erratum in: *American Journal of Human Genetics*, 2000, *66*, 2020.

Sham, P. C., Purcell, S., Cherny, S. S., & Abecasis, G. R. (2002). Powerful regression-based quantitative-trait linkage analysis of general pedigrees. *American Journal of Human Genetics*, *71*, 238–253.

Shephard, N. (1993). Maximum likelihood estimation of regression models with stochastic trend components. *Journal of the American Statistical Association*, *88*, 590–595.

Shephard, N. G., & Harvey, A. C. (1990). On the probability of estimating a deterministic component in the local level model. *Journal of Time Series Analysis*, *11*, 339–347.

Stram, D. O., & Lee, J. W. (1994). Variance components testing in the longitudinal mixed effects model. *Biometrics*, *50*, 1171–1177.

Stuart, A., & Ord, J. K. (1987). *Kendall's advanced theory of statistics. Volume 1. Distribution theory* (5th ed.). New York: Oxford University Press.

Thompson, R. (1973). The estimation of variance and covariance components with an application when records are subject to culling. *Biometrics*, *29*, 527–550.

Thompson, R. (2008). Estimation of quantitative genetic parameters. *Proceedings of the Royal Society of London. Series B. Biological Science*, *275*, 679–686.

Thompson, R., & Cameron, N. D. (1986). Estimation of genetic parameters. In: *3rd World Congress on Genetics Applied to Livestock Production, July 16–22, 1986, Lincoln, Nebraska* (pp. 371–381), Lincoln: University of Nebraska, Institute of Agriculture and Natural Resources.

Thompson, R., & Meyer, K. (1986). Estimation of variance components: What is missing in the EM algorithm? *Journal of Statistical Computation and Simulation*, *24*, 215–230.

Thompson, E. A., & Shaw, R. G. (1990). Pedigree analysis for quantitative traits: Variance components without matrix inversion. *Biometrics*, *46*, 399–413.

Thompson, E. A., & Shaw, R. G. (1992). Estimating polygenic models for multivariate data on large pedigrees. *Genetics*, *131*, 971–978.

Verbeke, G., & Molenberghs, G. (2003). The use of score tests for inference on variance components. *Biometrics*, *59*, 254–262.

Visscher, P. M. (2006). A note on the asymptotic distribution of likelihood ratio tests to test variance components. *Twin Research and Human Genetics*, *9*, 490–495.

Visscher, P. M., Andrew, T., & Nyholt, D. R. (2008). Genome-wide association studies of quantitative traits with related individuals: Little (power) lost but much to be gained. *European Journal of Human Genetics*, *16*, 387–390.

Visscher, P. M., & Duffy, D. L. (2007). The value of relatives with phenotypes but missing genotypes in association studies for quantitative traits. *Genetic Epidemiology*, *30*, 30–36 Erratum in: *Genetic Epidemiology*, 2007, *31*, 801.

Wald, A. (1943). Tests of statistical hypotheses concerning several parameters when the number of observations is large. *Transactions of the American Mathematical Society*, *54*, 426–482.

Williams, J. T., & Blangero, J. (1999a). Power of variance component linkage analysis to detect quantitative trait loci. *Annals of Human Genetics*, *63*, 545–563.

Williams, J. T., & Blangero, J. (1999b). Asymptotic power of likelihood ratio tests for detecting quantitative trait loci using the COGA data. *Genetic Epidemiology*, *17*, S397–S3402.

Wu, M. C., Lee, S., Cai, T., Li, Y., Boehnke, M., & Lin, X. (2011). Rare-variant association testing for sequencing data with the sequence kernel association test. *American Journal of Human Genetics*, *89*, 82–93.

Yang, J., Benyamin, B., McEvoy, B. P., Gordon, S., Henders, A. K., Nyholt, D. R., et al. (2010). Common SNPs explain a large proportion of the heritability for human height. *Nature Genetics*, *42*, 565–569.

Yang, J., Lee, S. H., Goddard, M. E., & Visscher, P. M. (2011). GCTA: A tool for genome-wide complex trait analysis. *American Journal of Human Genetics*, *88*, 76–82.

Yang, J., Manolio, T. A., Pasquale, L. R., Boerwinkle, E., Caporaso, N., Cunningham, J. M., et al. (2011). Genome partitioning of genetic variation for complex traits using common SNPs. *Nature Genetics*, *43*, 519–525.

Yang, J., Wray, N. R., & Visscher, P. M. (2010). Comparing apples and oranges: Equating the power of case-control and quantitative trait association studies. *Genetic Epidemiology*, *34*, 254–257.

CHAPTER TWO

Fungal Cell Wall Organization and Biosynthesis

Stephen J. Free[1]

Department of Biological Sciences, SUNY, University at Buffalo, Buffalo, New York, USA
[1]Corresponding author: e-mail address: free@buffalo.edu

Contents

Abstract

The composition and organization of the cell walls from *Saccharomyces cerevisiae*, *Candida albicans*, *Aspergillus fumigatus*, *Schizosaccharomyces pombe*, *Neurospora crassa*, and *Cryptococcus neoformans* are compared and contrasted. These cell walls contain chitin, chitosan, β-1,3-glucan, β-1,6-glucan, mixed β-1,3-/β-1,4-glucan, α-1,3-glucan, melanin, and glycoproteins as major constituents. A comparison of these cell walls shows that there is a great deal of variability in fungal cell wall composition and organization. However, in all cases, the cell wall components are cross-linked together to

Advances in Genetics, Volume 81
ISSN 0065-2660
http://dx.doi.org/10.1016/B978-0-12-407677-8.00002-6

33

generate a cell wall matrix. The biosynthesis and properties of each of the major cell wall components are discussed. The chitin and glucans are synthesized and extruded into the cell wall space by plasma membrane-associated chitin synthases and glucan synthases. The glycoproteins are synthesized by ER-associated ribosomes and pass through the canonical secretory pathway. Over half of the major cell wall proteins are modified by the addition of a glycosylphosphatidylinositol anchor. The cell wall glycoproteins are also modified by the addition of O-linked oligosaccharides, and their N-linked oligosaccharides are extensively modified during their passage through the secretory pathway. These cell wall glycoprotein posttranslational modifications are essential for cross-linking the proteins into the cell wall matrix. Cross-linking the cell wall components together is essential for cell wall integrity. The activities of four groups of cross-linking enzymes are discussed. Cell wall proteins function as cross-linking enzymes, structural elements, adhesins, and environmental stress sensors and protect the cell from environmental changes.

1. INTRODUCTION AND OVERVIEW: GENERAL ORGANIZATION OF THE FUNGAL CELL WALL

The cell wall is vital to the growth, survival, and morphogenesis of fungi. Mutational analysis has proved that it provides a protective barrier against a wide range of environmental conditions such as heat, cold, desiccation, and osmotic stress. It also provides protection against other microbes. Cell wall sensor proteins allow the fungus to assess and respond to changes in the environment. Cell wall adhesion and mucins mediate the adhesive properties of the fungal cell and play critical roles in allowing fungi to colonize new environments. The cell wall is also critical for participation in biofilm formation, a process that many fungi engage in, and is an important ecological niche for many fungi. For pathogenic fungi, the cell wall is critical for virulence and pathogenicity. The wall provides both adhesive properties critical for invasion of host tissue and protection against the host defense mechanisms. This review will focus on the cell walls of *Saccharomyces cerevisiae, Candida albicans, Aspergillus fumigatus, Neurospora crassa, Schizosaccharomyces pombe*, and *Cryptococcus neoformans*. These six species were chosen because their cell walls have been reasonably well characterized and they present a broad coverage of the fungi. Five of these species are ascomycetes and the sixth, *C. neoformans*, is a basidiomycete. *C. neoformans* is unique among these fungi in that it produces a large polysaccharide capsule exterior to the cell wall. This capsule is a major virulence factor and has been extensively characterized (Doering, 2009).

Pathogenic fungi share a large amount of cell biology with their plant and animal hosts. Because fungal pathogens share a common biology with their hosts, finding antifungal agents for treatment of fungal infections is a daunting task. However, the fungal cell wall is unique to the pathogen and has therefore become a favorite target for the development of antifungal agents. Along with fungicides targeting ergosterol function, fungicides targeting cell wall biogenesis are among the most effective antifungal agents currently used to treat fungal infections.

Several good reviews on the fungal cell wall have appeared in the literature over the years (Bowman & Free, 2006; Chaffin, 2008; Klis, Boorsma, & de Groot, 2006; Latgé, 2007; Latgé et al., 2005; Lesage & Bussey, 2006; Ruiz-Herrera, Elorza, Valentin, & Sentandreleu, 2006). The intents of this review are to give an update on the most recent literature, examine cell wall structures and functions from a geneticist's viewpoint, and to highlight some of the diversity that exists in fungal cell walls. This diversity exists not only between the cell walls produced by different fungal species but can also be seen in the cell walls of different cell types produced by a single fungal species. There is a large literature associated with the fungal cell wall and the author apologizes for inadvertently omitting any publications that might have been referenced in the review, but were not.

Fungal cell walls are composed of glucans, chitin and chitosan, mannans and/or galactomannans, and glycoproteins. A variety of glucans, including β-1,3-, mixed β-1,3-/β-1,4-, β-1,6-, and α-1,3-glucans, have been identified in fungal cell walls. These glucans are synthesized by plasma membrane-associated glucan synthases, which extrude newly formed linear glucan polymers through the plasma membrane into the cell wall space. The chitin is synthesized by plasma membrane-associated chitin synthases and also extruded into the cell wall space as a linear polymer. The fungi have multiple chitin synthases which provide for a redundancy in chitin synthetic function. In some cases, different chitin synthases have been shown to function in different cell types or during different times in the cell cycle (Roncero, 2002; Ruiz-Herrera, González-Prieto, & Ruiz-Medrano, 2002). The extruded chitin can be acted on by chitin deacetylases to generate chitosan, a deacetylated chitin polymer.

Cell wall glycoproteins are synthesized by ER-associated ribosomes and cotranslationally extruded into the lumen of the ER. As the cell wall glycoproteins enter the lumen of the ER, N-linked oligosaccharides are added to asparagine residues found in the primary amino acid context of asparagine-X-serine or threonine where X can be any amino acid except proline.

Many cell wall proteins contain a GPI (glycosylphosphatidylinositol) anchor. GPI anchor addition occurs in the ER within seconds after the cell wall proteins are released into the lumen of the ER. As the cell wall proteins pass through the ER and Golgi apparatus, O-linked glycosylation occurs (addition of mannans or galactomannan) and the N-linked oligosaccharides are further processed to generate high mannan outer chain structures (in *S. cerevisiae, C. albicans*) or N-linked galactomannans (*S. pombe, A. fumigatus, N. crassa*). Lipid-linked galactomannans may also be generated in the compartments of the secretory pathway (Costachel, Coddeville, Latgé, & Fontaine, 2005; Latgé, 2009). The glycoproteins (and presumably the lipid-linked galactomannans) are secreted into the cell wall space by vesicular trafficking. Once the glucans, chitin, and glycoproteins are released into the cell wall space, cell wall cross-linking enzymes covalently join the various cell wall components together to generate a three-dimensional chitin/glucan/glycoprotein matrix. These cross-linking enzymes recognize the glucans, chitin, and the oligosaccharide polymers found as posttranslational modifications on the glycoproteins and function as transglycosidases. They have a glycosyl hydrolase function to cut one polymer and a glycosyl transferase function to create a new bond that cross-links one polymer to another. EM microscopy images of cell walls suggest the region next to the plasma membrane is rich in carbohydrate, while the outer portion of the cell wall is rich in protein. However, it is likely that many GPI-anchored proteins may traverse the carbohydrate-rich inner layer of the cell wall. Figure 2.1 gives a schematic representation of the basic fungal cell wall organization.

The roles for each of the individual cell wall components (glucans, chitin and chitosan, mannans, galactomannans, and glycoproteins) can be assessed by examining mutants affected in their biosynthesis. Loss of any of these cell wall components can dramatically affect the growth, morphology, and viability of the fungus, demonstrating that all of these components are needed for the formation of a normally functional cell wall. In this review, the biosynthesis of each of these cell wall components will be examined, but first we will briefly review some of the approaches and techniques used to study fungal cell walls.

2. METHODS USED IN STUDYING THE CELL WALL

Because the cell wall components are all cross-linked together and the proteins are heavily glycosylated, it is challenging to study the organization and components of the fungal cell wall. However, genetic, microscopic,

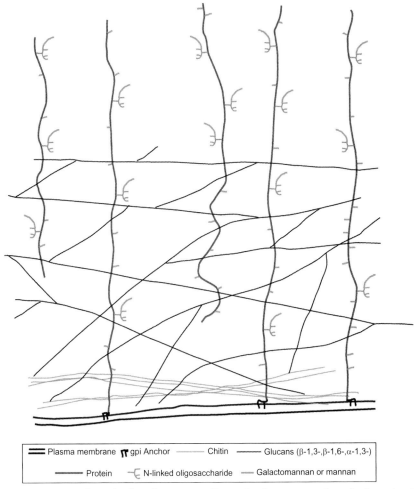

Figure 2.1 Basic fungal cell wall structure. Generalized representation of a fungal cell wall. The chitin and glucan matrix are cross-linked together to form a glucan/chitin matrix. The glucan fraction contains β-1,3-glucan as a major element and can include other types of glucans, such as β-1,6-glucan and α-11,3-glucan. GPI-anchored glycoproteins and nonanchored glycoproteins are covalently attached to the cell wall. These proteins contain N-linked oligosaccharides that have been modified by the addition of either galactomannan or mannans. Attached mannans can contain up to 200 mannose residues, while attached galactomannans are much smaller. The glycoproteins also contain short O-linked galactomannans or mannans. (See Color Insert.)

2D-gel electrophoresis, proteomic, and immunological approaches have allowed for the identification of cell wall proteins, and the major polymers present in cell walls have been identified by carbohydrate analyses (Klis et al., 2006; Latgé, 2007; Pitarch, Sánchez, Nombela, & Gil, 2002). EM studies have provided critical information about the location of different components within the wall, giving a picture of a carbohydrate-rich inner cell wall layer and a protein-rich outer layer. Some cell wall proteins can be released from some fungal cell walls by treatments with HF-pyridine and alkali, and these treatments were used to release and study cell wall proteins (De Groot et al., 2004). Other cell wall proteins have to be released by treatments with glucanases and chitinases.

With the development of mass spectrometry techniques for protein identification, the sequencing of fungal genomes, and the development of bioinformatics, it has become much easier to identify cell wall proteins and characterize their functions. For example, many cell wall proteins are GPI-anchored, and programs to identify GPI-anchored proteins in sequenced fungal genomes have been used to identify a number of prospective cell wall proteins (De Groot, Hellingwerf, & Klis, 2003; De Groot et al., 2009; Eisenhaber, Schneider, Wildpaner, & Eisenhaber, 2004; Plaine et al., 2008). The development of methods to identify proteins with a mass spectrometry analysis has been particularly helpful in identifying cell wall proteins. A number of cell wall proteins have been identified by mass spectrometry in *S. cerevisiae, C. albicans, A. fumigatus, N. crassa, S. pombe*, and *C. neoformans* (Bruneau et al., 2001, De Groot et al., 2007; Eigenheer, Lee, Blumwald, Phinney, & Gelli, 2007; Maddi, Bowman, & Free, 2009; Yin, de Groot, de Koster, & Klis, 2007). In some cases, the cell wall proteins have been released from the cell wall matrix by glucanase and chitinase treatments, trypsinized, and the tryptic fragments identified by mass spectrometry. In other cases, a "trypsin shaving" approach was used. In these studies, purified cell walls or living cells were directly treated with trypsin and the released tryptic fragments were collected and identified by mass spectrometric analysis. Recently, trifluoromethanesulfonic acid, which specifically cleaves glycosidic bonds and leaves peptide bonds intact, has been used to treat fungal cell walls and release the cell wall proteins, which were then trypsinized and the released peptides analyzed by nano-LC/MS/MS (Maddi et al., 2009). The trifluoromethanesulfonic acid approach has the advantage of allowing the investigator to identify sites of N-linked oligosaccharide addition. The first glucosamine from the oligosaccharide attached to the aspargine residue is not released by trifluoromethanesulfonic acid, and tryptic peptides can be identified with the attached glucosamine (Birkaya, Maddi, Joshi, Free, & Cullen, 2009; Maddi et al., 2009). The sites of O-linked oligosaccharide attachment can also be determined by subjecting cell wall proteins to a β-elimination using NH_4OH, which removes the oligosaccharide and tags the site with NH_3 (Rademaker et al., 1998). Following the β-elimination, the protein is then subjected to trypsinization and mass spectrometric analysis to identify the peptides containing the NH_3 tag.

As a result of these recent studies identifying cell wall proteins with mass spectrometric analysis, a group of proteins that are commonly found in fungal cell walls can be identified. Table 2.1 identifies those commonly found cell wall proteins and shows whether the proteins have been identified in each of the five well-characterized ascomycetes. Homologs of these commonly found cell wall proteins are encoded in almost all sequenced fungal genomes, indicating that they have been evolutionarily conserved, and suggesting that they perform important functions. These different cell wall proteins and their functions will be a major focus of this review.

3. THE SPECIFIC COMPONENTS FOUND IN FUNGAL CELL WALLS

3.1. Chitin and chitosan

Chitin makes up between 1% and 15% of the cell wall mass, with yeast species generally having 1–2% chitin and filamentous fungi having up to 15% chitin (Table 2.2). Chitin is synthesized by plasma membrane-associated chitin synthases. These enzymes utilize UDP-N-acetylglucosamine from the cytoplasmic side of the membrane as a substrate to synthesize linear chitin molecules (polymers of β-1,4-N-acetylglucosamine). The chitin polymer is elongated by the addition of N-acetylglucosamine to the nonreducing end of the polymer and is extruded into the cell wall space, reducing end first, during synthesis. The chitin synthases have multiple transmembrane domains that form a channel through which the elongating chitin polymer is passed. Chitin polymers in the cell wall space can participate in interchain hydrogen bonding and assemble into microfibrils in which the individual chitin molecules are arranged in an antiparallel pattern (Ruiz-Herrera et al., 2006). These microfibrils have enormous tensile strength and significantly contribute to the rigidity and integrity of the cell wall.

There is a large literature available on the synthesis of chitin. All characterized fungi have chitin synthases and most have multiple genes encoding chitin synthases. *S. cerevisiae* has three genes that encode chitin synthases

Table 2.1 Major glycoproteins that are commonly found in fungal cell walls

Proteins	S. cerevisiae	C. albicans	A. fumigatus	N. crassa	S. pombe
Endochitinases	Cts1p	Cht2p	–	Chit-1	–
Gycosylhydrolase family 72	Gas1p, Gas3p, Gas5p	Phr1p, Phr2p, Phr4p, Pga4p	GEL1, GEL4	Gel-1, Gel-2, Gel-5	Gas1, Gas5
Gycosylhydrolase family 17	Bgl2p, Scw4p, Scw10p	Bgl2p, Mp65p, Scw4p	–	GH17-3 NCU09175, GH17-4 NCU09326	–
Gycosylhydrolase family 16	Crh1p, Crh2p	Crh11, Crh12	Crh1, Crh2	GH16-1 NCU01353, GH16-7 NCU05974	–
EM33/ACW-1	Ecm33p	Ecm33	Ecm33	ACW-1	Ecm33
Pir proteins	Pir1p, Pir2p, Pir3p, Pir4p	Pir1	–	–	–
Adhesins and flocculins	Flo11p, Flo10p	Als1p, Als3p, Als4p, Rbt1p, Rhd3p/ Pga29p, Pga24p, Hwp1p, Flo1p/Pga62p	–	–	–
Catalases, superoxide dismutase	–	Sod4/Pga2, Sod5, Sod6/Pga9	–	CAT-3, ACW-10	–

Proteins identified by proteomic analyses of cell wall proteins are listed. Those proteins given in bold lettering are GPI-anchored proteins and those given in normal type do not have a GPI anchor. Proteomic analyses from S. cerevisiae (Cappellaro, Mrsa, & Tanner, 1998; Pardo et al., 2000; Yin et al., 2005; Birkaya et al., 2009; Insenser et al., 2010), C. albicans (De Groot et al., 2004; Castillo et al., 2008; Klis, Sosinska, de Groot, & Brul, 2009; Maddi et al., 2009; Heilmann et al., 2011; Sosinska et al., 2011), A. fumigatus (Bruneau et al., 2001), N. crassa (Bowman, Piwowar, Al Dabbous, Vierula, & Free, 2006; Maddi et al., 2009; Maddi & Free, 2010), and from S. pombe (De Groot et al., 2007) were used in constructing the table.

Table 2.2 Cell wall components and their presence in different fungal species

Cell wall component	S. cerevisiae	C. albicans	A. fumigatus	N. crassa	S. pombe	C. neoformans
Chitin	Yes (1–2%)	Yes (2–6%)	Yes (7–15%)	Yes (4%)	Not in vegetative cell (found in conidia)	Yes
β-1,3-Glucan	Yes (50–55%)	Yes (30–39%)	Yes (20–35%)	Yes (87%)	Yes (46–54%)	Yes
Mixed β-1,3/1,4-glucan	No	No	Yes	Possibly	No	No
β-1,6-Glucan	Yes (10–15%)	Yes (43–53%)	No	No	No	Yes
α-1,3-Glucan	No	No	Yes (35–46%)	Only in conidial cell wall	Yes (18–28%)	Yes—anchors capsule to the wall
Outer chain mannan	Yes (10–20%)	Yes (38–40%)	No	No	No	No
Galacto-mannan	No	No	Yes—20–25% with a tetrasaccharide repeat	Yes (12%)	Yes (9–14%)	Yes
Melanin	No	Yes—during infection	Yes—in conidia	Yes—in the ascospore and perithecium	No	Yes—during infection

The cell wall components listed were identified from the following fungi: *S. cerevisiae* (Lesage & Bussey, 2006), *C. albicans* (Ruiz-Herrera et al., 2006), *A. fumigatus* (Gastebois, Clavaud, Aimanianda, & Latgé, 2009), *N. crassa* (Maddi & Free, 2010), *S. pombe* (Matsuo, Tanaka, Nakagawa, Matsuda, & Kawamukai, 2004; Magnelli, Cipollo, & Robbins, 2005), and *C. neoformans* (Doering, 2009).

(Cabib, Roh, Schmidt, Crotti, & Varma, 2001), *C. albicans* has four (Munro et al., 2003), *A. fumigatus* has seven (Mellado et al., 2003), *C. neoformans* has eight (Banks et al., 2005; Klutts, Yoneda, Reilly, Bose, & Doering, 2006), and *N. crassa* has seven (Borkovich et al., 2004). *S. pombe* has two chitin synthase genes (Matsuo et al., 2004) but one of these has been reported to lack synthase activity (Martin-Garcia, Durán, & Valdivieso, 2003). In most of these fungi, one or two of the chitin synthases have been found to be responsible for the majority of the chitin synthesized by the vegetative hyphae, with other chitin synthases producing lesser amounts (Beth-din & Yarden, 2000; Mellado et al., 2003; Mio et al., 1996; Munro et al., 2001). The trafficking and activation aspects of chitin synthases have also been actively researched. It is clear that the activation and targeting of chitin synthases to the hyphal tip and septum are tightly regulated in filamentous fungi (Lenardon et al., 2010; Sánchez-León et al., 2011). The presence of multiple chitin synthases in many fungi suggests that different chitin synthases might be used for chitin production at different stages of the fungal life cycles or for particular cell types and the available research strongly supports this hypothesis. Different chitin synthases have been shown to be differentially involved in septum formation, hyphal growth, and development (Horiuchi, 2009; Lee et al., 2004; Roncero, 2002). Mutational analyses in *S. cerevisiae*, *C. albicans*, *A. fumigatus*, and *N. crassa* show that the loss of the major chitin synthase(s) results in a weakened cell wall, a reduced growth rate, and an increase in hyphal lysis. However, the analysis of chitin synthesis in *S. pombe* presents an interesting contrast. Chitin was not detected in an analysis of the *S. pombe* vegetative cell wall (De Groot et al., 2007; Magnelli et al., 2005) and deletion of the chitin synthase genes, chs1 and chs2, does not affect vegetative growth. However, the chs1 gene is required for the development of asci, and the chs2 gene is involved in septum formation (Arellano, Cartagena-Lirola, Hajibagheri, Durán, & Valdivieso, 2000; Matsuo et al., 2004). *S. pombe* demonstrates that chitin is not absolutely required for all fungal cell walls and that chitin can be a cell-type-specific cell wall component.

Because chitin is a critical cell wall component for pathogenic fungi, chitin synthesis has been considered a prime target for the development of antifungal agents. The best-known chitin synthesis inhibitors are the naturally occurring nikkomycins and polyoxins. These compounds are analogs of UDP-*N*-acetylglucosamine, the substrate for chitin synthesis, and act as competitive inhibitors for chitin synthases (Ruiz-Herera & San-Blas, 2003). Unfortunately, nikkomycins and polyoxins have not proved to be effective in controlling fungal infections, probably because of their limited uptake by the fungi. However, they are currently being used in conjunction with other antifungal agents to treat fungal diseases.

Most cell wall chitin functions as a major structural component of the cell wall. However, in some fungi the extruded chitin polymers can be acted upon by chitin deacetylases to generate chitosan, a polymer of glucosamine residues. Chitosan is much more soluble than chitin and much less is known about the role of chitosan. The amount of chitin that is converted to chitosan is likely to be different for the various fungal species and may also vary with cell type. *S. cerevisiae* has two sporulation-specific chitin deacetylase genes, and mutational analysis shows that the spore cell wall from the double mutant has an increased sensitivity to lysis, caused by digestive enzyme treatment, suggesting that the chitosan provides some protection to the spore (Christodoulidou, Bouriotis, & Thireos, 1996). For *S. cerevisiae*, chitosan is a cell-type-specific cell wall component. A mutational analysis on the role of chitosan in *C. neoformans* showed that mutants lacking chitin deacetylase activity had a decreased rate of growth and reduced pathogenicity (Baker, Specht, & Lodge, 2011). This demonstrates that the formation of chitosan from preexisting chitin can be an important step in cell wall biogenesis for some fungi.

3.2. β-1,3-Glucans

β-1,3-Glucan is a major constituent of all of the characterized fungal cell walls, making up between 30% and 80% of the mass of the wall (Table 2.2). In the cell wall, β-1,3-glucan is found as a branched polymer, with the branches being attached to the core polymer by β-1,6-branches (Klis et al., 2006, Latgé, 2007; Lesage & Bussey, 2006). A chemical analysis of purified β-1,3-glucans indicates that they can exist in a single-stranded helical conformation in solution, and that they can also form stable three-stranded helices (Bohn & BeMiller, 1995; Laroche & Michaud, 2007). The β-1,3-glucan helix is thought to function as a coiled spring-like structure to confer a degree of elasticity and tensile strength to the cell wall (Lesage & Bussey, 2006). Thus, β-1,3-glucan is well suited to function as the major cell wall building block. The glucan is synthesized by β-1,3-glucan synthase, a well-characterized plasma membrane-associated enzyme with multiple transmembrane domains (Douglas et al., 1994; Klis et al., 2006; Lesage & Bussey, 2006; Thompson et al., 1999). The enzyme utilizes cytoplasmic UDP-glucose as a substrate and adds glucose residues to the growing linear glucan polymer (Frost, Brandt, Capobianco, & Goldman, 1994). As was the case for the chitin synthases, the β-1,3-glucan adds glucose to the nonreducing end of the polymer. The polymer is extruded into the cell wall space through a channel formed by the transmembrane domains during polymer synthesis. *A. fumigatus, C. neoformans, N. crassa,* and *C. albicans* have a single β-1,3-glucan synthase, called FKS1 (Beauvais et al., 2001; Tentler et al., 1997; Thompson et al., 1999). *S. cerevisiae* has three glucan synthase genes, but FKS1 functions as the major β-1,3-glucan synthase during vegetative growth in yeast (Klis et al., 2006; Lesage & Bussey, 2006). *S. pombe* has four β-1,3-glucan synthase genes, one of which has been shown to be involved in septation, polarized growth, mating, spore formation, and spore germination (Cortés, Ishiguro, Durán, & Ribas, 2002; Liu, Wang, McCollum, & Balasubramanian, 1999). Mutants lacking a functional FKS1 are dramatically affected in their morphology and growth (Klis et al., 2006; Lesage & Bussey, 2006; Liu et al., 1999; Tentler et al., 1997, Thompson et al., 1999).

The localization and targeting of FKS1 has been followed in *S. cerevisiae, N. crassa,* and *S. pombe*. In *S. cerevisiae,* a GFP-tagged FKS1 was shown to localize to areas of polarized growth and to colocalize with actin patches (Utsugi et al., 2002). In *S. pombe,* FKS1 has been shown to localize to the forming septum (Cortés et al., 2002). In *N. crassa,* FKS1 has been shown to be delivered through the Spitzenkörper to the growing hyphal tip (Verdín, Bartnicki-Garcia, & Riquelme, 2009). Interestingly, the vesicles carrying FKS1 and the vesicles carrying chitin synthase were shown to be two different vesicle populations and to be located in two spatially distinct areas within the Spitzenkörper (Verdín et al., 2009).

FKS1 is found in association with the RHO1 G-protein, which functions as a regulatory subunit for FKS1 (Klis et al., 2006; Lesage & Bussey, 2006; Mazur & Baginsky, 1996). RHO1 functions in the signal transduction systems that control the MAP kinase signal pathways regulating fungal growth and the cell wall integrity (CWI) response. RHO1 association is required to activate glucan synthesis by FKS1. This requirement for RHO1 activation of glucan synthesis allows the cell to produce β-1,3-glucan when it is needed for growth or to strengthen the cell wall in response to cell wall stress, and to curtail glucan synthesis when it is not needed.

Because of its importance to cell wall biogenesis, β-1,3-glucan synthase is a prime target for the development of antifungal agents. The echinocandin family of β-1,3-glucan synthase inhibitors, caspofungin, micafungin, and anidulafungin, have found extensive clinical usage (Chen, Slavin, &

Sorrell, 2011). They are noncompetitive inhibitors of β-1,3-glucan synthase and treatment with the echinocandins results in cell swelling and lysis. The echinocandins are currently being used in clinical settings for the treatment of aspergillosis and candidiasis.

3.3. Mixed β-1,3-/β-1,4-glucans

Studies in *A. fumigatus* demonstrated the existence of a mixed linkage glucan polymer with both β-1,3- and β-1,4-linkages (Fontaine et al., 2000). The polymerase was found to be a major constituent of the cell wall. Compositional studies of the *N. crassa* cell wall also demonstrated the presence of a large fraction of 1,4-linked glucose residues, suggesting that it might also contain such a polymer (Maddi & Free, 2010). No genes encoding a mixed-glucan synthase have been identified and virtually nothing is known about how this polymer is synthesized or whether the polymer is restricted to the cell walls of particular cell types.

3.4. β-1,6-Glucans

β-1,6-Glucan has been shown to be an important component of the *S. cerevisiae* and *C. albicans* cell walls (Aimanianda et al., 2009; Klis et al., 2006; Lesage & Bussey, 2006) and has also been shown to be a major constituent of the *C. neoformans* cell wall (James, Chemiak, Jones, Stortz, & Reiss, 1990) (Table 2.2). In *S. cerevisiae*, β-1,6-glucan has been shown to form cross-links with β-1,3-glucan, chitin, and with the GPI anchor oligosaccharide (Kapteyn et al., 1996; Kollár, Petráková, Ashwell, Robbins, & Cabib, 1995; Kollár et al., 1997; Lu et al., 1995). Cross-linking the GPI anchor to the cell wall matrix covalently attaches the GPI-anchored cell wall proteins into the cell wall. Thus, the β-1,6-glucan plays an important role in the formation of the *S. cerevisiae* cell wall matrix (Klis et al., 2006; Lesage & Bussey, 2006).

The β-1,6-glucan synthase has yet to be definitively identified in any fungal species. However, a number of genes that affect the synthesis of β-1,6-glucan have been identified in *S. cerevisiae* (Lesage & Bussey, 2006). These include the KRE5, BIG1, and ROT1 proteins, which have been localized to the ER and which are likely to function in the quality control/protein-folding activities that control the exit of proteins from the ER to the Golgi apparatus. The KRE6 and SKN1 proteins are also needed for normal β-1,6-glucan synthesis. Sequence and structural considerations suggest that these proteins are glycosyl hydrolases or transglycosidases and they might function to cross-link the β-1,6-glucan into the cell wall. However, protein localization studies indicate that KRE6 and SKN1 are localized to the Golgi apparatus, a location that would be inconsistent with a role in cross-linking β-1,6-glucan into the cell wall matrix. KRE1 is a GPI-anchored protein that is required for normal levels of cell wall β-1,6-glucan. The analysis of the KRE1 mutant cell wall shows that the β-1,6-glucan is shorter than normal and suggests KRE1 functions in the cell wall space to elongate (cross-link together) β-1,6-glucans (Boone, Sommer, Hensel, & Bussey, 1990). KRE9 and KNH1 are a duplicate pair of non-GPI-anchored cell wall proteins that are likely to function in cross-linking β-1,6-glucan into the wall (Lesage & Bussey, 2006). Loss of both KRE9 and KNH1 is lethal. These enzymes, which have been genetically identified as being needed for the synthesis of normal levels of β-1,6-glucan, are likely to be necessary for the proper folding and targeting of the yet to be identified β-1,6-synthase to the plasma membrane or for the cross-linking of the polymer to cell wall components. The site of β-1,6-glucan synthesis has been shown to be at the plasma membrane (Montijn et al., 1999), suggesting that the β-1,6-synthase, when identified, will function similar to the chitin synthases and β-1,3-glucan synthases. With the availability of the *S. cerevisiae* mutant libraries and the other tools available for yeast, it is likely that we will soon be able to identify the β-1,6-glucan synthase.

A recent study showed that many of the genes identified in *S. cerevisiae* as being important for β-1,6-glucan synthesis are also important for β-1,6-glucan synthesis in *C. neoformans* (Gilbert et al., 2010). *C. neoformans* mutants in the genes encoding KRE5, KRE6, and SKN produced reduced amounts of β-1,6-glucan, and had complex phenotypes, which included altered morphologies, sensitivity to environmental stress, and a reduction in the exopolysaccharide capsule. These mutants were avirulent in the mouse inhalation model of infection. These findings demonstrate that β-1,6-glucan is essential to the *C. neoformans* cell wall.

While β-1,6-glucans have been found in *S. cerevisiae, C. albicans*, and *C. neoformans*, they have not been found in the analyses of other fungal cell walls (*S. pombe, A. fumigatus*, and *N. crassa*), and genes encoding the proteins which have been identified in *S. cerevisiae* affecting β-1,6-glucan production are not found in the genomes of these other fungi. This indicates that the involvement of β-1,6-glucan in the cross-linking of chitin, β-1,3-glucan, and glycoprotein into the cell wall matrix, is restricted to a limited number of fungal species. *S. pombe, A. fumigatus*, and *N. crassa* must use other polymers for cross-linking their cell wall components together.

3.5. α-1,3-Glucans

α-1,3-Glucans have been found in a number of fungal cell walls. The α-1,3-glucan synthases have a predicted structure with a cytoplasmic synthase domain that would use UDP-glucose as a substrate to add glucose residues to the nonreducing end of the glucan, and multiple transmembrane domains that could form a channel through which the reducing end of the polymer could be extruded across the plasma membrane into the cell wall space. Interestingly, the α-1,3-glucan synthases also have a large extracellular N-terminal region with homology to the transglucanases, which could function in cross-linking newly synthesized α-1,3-glucan to other cell wall components. The α-1,3-glucan synthase gene (*ags1*) has been shown to be essential for *S. pombe* (Grün et al., 2004; Hochstenbach et al., 1998). In a *S. pombe* temperature-sensitive *ags1* mutant, cell lysis occurs at the restrictive temperature, demonstrating the importance of α-1,3-glucan for the integrity of the *S. pombe* cell wall. In *C. neoformans*, cell wall α-1,3-glucan anchors the polysaccharide capsule to the cell wall. *C. neoformans ags1* mutants release capsule polysaccharides into the medium and are avirulent (Reese et al., 2007). In contrast to the situation in *C. neoformans*, a decrease of α-1,3-glucan from the *A. fumigatus* cell wall results in an increase in virulence (Beauvais et al., 2001; Fontaine et al., 2010; Maubon et al., 2006). *A. fumigatus* has three α-1,3-glucan synthases, AGS1, AGS2, and AGS3, and two of these have been shown to participate in the synthesis of cell wall α-1,3-glucan. In *N. crassa*, α-1,3-glucan can be immunologically detected in the conidial cell wall, but not in the cell wall of vegetative hyphae. Loss of the *N. crassa* α-1,3-glucan synthase, AGS-1, disrupts conidial differentiation (personal observation). Thus, in *N. crassa*, α-1,3-glucan seems to be a cell-type-specific cell wall component and highlights the diversity in cell wall structure that can exist in different cell types within a single fungal species. The available data indicates that α-1,3-glucans are absent from some fungal cell walls (*S. cerevisiae, C. albicans*), play an auxiliary role in the vegetative cell walls of other fungi (*A. fumigatus*), are essential for the vegetative cell walls of a third group of fungi (*S. pombe, C. neoformans*), and represents a cell-type-specific cell wall component for yet other group fungi (*N. crassa*).

3.6. Melanins

Melanins are frequently found in the cell walls of fungal conidia, ascospores, and in other structures that are exposed to harsh environments (Eisenman & Casadevall, 2012). Melanin is a polymer formed from the cross-linking of phenolics. At least two pathways are known to occur in fungi for the formation of melanin. One of these pathways utilizes tyrosinase and/or laccase to generate DOPA (L-3,4-dihydroxyphenylalanine) from tyrosine. The DOPA is then polymerized to form melanin. The second pathway has 1,8-dihydroxynapthalene (DHN), instead of DOPA as the important intermediate in melanin formation. The DHN pathway utilizes polyketide synthases and laccases in a melanin biosynthetic pathway. Melanin is related to lignin, a component of wood, and functions in the fungal cell wall in an analogous fashion to harden the wall and to protect the wall from UV irradiation. During melanin synthesis, the covalent bonds joining the individual phenolics are randomly formed via a free-radical reaction and the individual phenolics within the polymerized molecule are linked together by a number of the different bonds. As a consequence of having their individual phenolic components joined together by several different linkages and having these bonds formed in a random manner, melanin and lignin are structurally diverse and complex, and are difficult to digest enzymatically.

The pathogenic fungus *A. fumigatus* uses the DHN pathway to produce melanized conidia, and the genes encoding the polyketide synthase and other enzymes needed for the synthesis of DHN have been identified in the *A. fumigatus* genome (Langfelder et al., 1998; Sugareva et al., 2006; Tsai, Wheeler, Chang, & Kwon-Chung, 1999). Conidia produced during the infection process contained melanized cell walls (Youngchim, Morris-Jones, Haly, & Hamilton, 2004), and the presence of melanized conidial cell wall is an important virulence factor for *A. fumigatus* (Karkowska-Kuleta, Rapala-Kozik, & Kozik, 2009; Tsai et al., 1999). Melanin is also an important virulence factor for *C. neoformans* (Jacobson, 2000; Karkowska-Kuleta et al., 2009; Wang, Aisen, & Casadevall, 1995; Williamson, 1997). During infection, *C. neoformans* obtains the phenolics needed for the DOPA pathway of melanin synthesis from the host tissues and uses a cell wall laccase, LAC1, to facilitate melanin formation (Eisenman & Casadevall, 2012). *C. albicans*, like *C. neoformans*, uses the DOPA pathway to form cell wall melanin from host phenolics (Morris-Jones et al., 2005; Walker et al., 2010). Melanization of the *C. albicans* cell wall occurs during the infection process (Morris-Jones et al., 2005). However, the importance of the melanin as a virulence factor for *C. albicans* remains to be firmly established. Melanin could function as a virulence factor on several levels (Jacobson, 2000). It is a strong antioxidant and could absorb oxygen free-radicals that are produced by the host. It also strengthens the cell wall and makes it much more resistant to degradation by the host. It helps protect the cell from lysis that might occur because of treatments with antifungal agents, which would otherwise weaken the cell wall. Melanin, which strongly absorbs UV light, also provides protection against UV light damage for many fungi (Eisenman & Casadevall, 2012).

N. crassa is not pathogenic, but melanin is an important cell wall component for the *N. crassa* ascospore and perithecium (female mating structure). *N. crassa* utilizes the DOPA pathway to produce melanin in the cell walls of those cells on the outer surface of the perithecium. Mutants lacking tyrosinase, which is used to generate the DOPA for melanin biosynthesis, are not able to produce functional perithecia (Fuentes, Connerton, & Free, 1994; Lerch, 1983). The cell wall of the *N. crassa* ascospore is highly melanized, and the melanization is thought to be an important protectant for the ascospore. The literature indicates that melanin is an important cell wall polymer for some cells, but that it is a cell-type-specific cell wall component. It is particularly important for strengthening the wall under conditions of pathogenesis or for cells that need to survive in harsh environments.

3.7. Cell wall glycoproteins

A number of glycoproteins have been found associated with the cell wall (see Table 2.1 for a list of major integral cell wall proteins found in the cell walls from *S. cerevisiae, C. albicans, A. fumigatus, N. crassa* and *S. pombe*). Some glycoproteins are covalently linked into the cell wall matrix and might be called "integral cell wall proteins." These proteins will be highlighted and discussed below. In addition to these integral proteins, a large number of additional "nonintegral" proteins are found associated with the wall and may have important functions. The hydrophobins, which are found on conidial cell walls, are examples of such proteins.

Many of the integral cell wall proteins are produced as GPI-anchored proteins, but there are also a large number of non-GPI-anchored proteins which are incorporated into the cell wall matrix. These cell wall proteins can also be grouped according to their functions. After discussing the GPI-anchoring process and the addition of mannan and galactomannan posttranslational modifications, the integral cell wall proteins will be discussed according to their functions. Many of the integral proteins function in cross-linking the matrix together and are important for cell wall biogenesis. Other cell wall proteins have been shown to function as cell wall sensors, as adhesins, in iron acquisition, and in oxygen free-radical detoxification. Last, there are cell wall proteins to which no activity can be ascribed. These are normally called "structural" proteins, insinuating that they function in creating the structure of the wall. Some of these structural proteins may be assigned other functions as we gain a better understanding of biofilms and other activities in which the cell wall participates. All of the proteins mentioned above share the properties of having an N-terminal signal peptide and traversing the secretory pathway before being delivered to the cell wall space.

Investigators characterizing the cell wall invariably find a common set of "cytosolic" proteins in their cell wall preparations, even when the preparations have been rigorously treated with alkali, high salts, and boiling SDS solutions. Enolase, glyceraldehyde-3-phosphate dehydrogenase, pyruvate decarboxylase, alcohol dehydrogenase, translation elongation factor 1α, and heat shock protein SSA1 are examples of cytosolic proteins that are typically found in cell wall preparations (Castillo et al., 2008; De Groot et al., 2004; Ebanks, Chisholm, McKinnon, Whiteway, & Pinto, 2006; Maddi et al., 2009; Pardo et al., 2000). These proteins have been found to be secreted by regenerating *S. cerevisiae* protoplasts (Pardo et al., 2000), which suggests they are released by cells, and that their presence in the cell wall is not due to contamination of the cell wall by cytosolic proteins during cell lysis and the isolation of the cell wall. These proteins are often referred to as "noncanonical" or "nonconventional" cell wall proteins. Whether these noncanonical proteins, which are cross-linked into the cell wall matrix, should be considered cell wall proteins, how they get incorporated into the wall, and whether they have cell wall-specific functions, has been enigmatic. Most treatises on the cell wall largely disregard these proteins.

3.7.1 GPI anchoring
Many cell wall proteins are synthesized with a GPI anchor (see Table 2.1). Only proteins passing through the secretory pathway (proteins with an N-terminal signal peptide) are subject to GPI-anchoring. The GPI anchor is a glycolipid structure that functions to anchor proteins into the outer leaflet of the plasma membrane (and to the luminal leaflet of the ER and Golgi apparatus while the protein is in transit to the plasma membrane) (Ferguson, Kinoshita, & Hart, 2009). Proteins are targeted for GPI anchor addition by an amino acid sequence motif found at the C-terminus of GPI-anchored protein. GPI-anchored proteins contain a region of between 8 and 18 hydrophobic amino acids at their extreme C-terminus, which may hold the newly synthesized protein in the luminal leaflet of the ER membrane when the protein is released from the translocon. Immediately preceding the hydrophobic amino acid sequence is a region with a small amino acid

called the omega site (glycine, alanine, and serine are commonly used amino acids), which is flanked by short regions of disordered amino acid sequence (regions that do not form strong secondary structure). During GPI anchor addition, the omega site is cleaved and the protein is attached to an ethanolamine residue in the GPI anchor.

The GPI anchor is built on the ER membrane in a process similar to the creation of the N-linked oligosaccharide. A review by Eisenhaber, Maurer-Stroh, Novatchkova, Schneider, and Eisenhaber (2003) provides detailed information about the GPI anchor structure and the enzymes involved in its synthesis. A number of enzymes act sequentially to add an N-acetylglucosamine, mannose residues, and a phosphoethanolamine onto a phosphatidylinositol lipid. All GPI anchors have a core structure consisting of a phosphatidylinositol to which is attached an oligosaccharide chain terminated by a phosphoethanolamine. The oligosaccharide consists of a glucosamine residue (initially added as GlcNAC and then deacetylated) attached by an α-1,6-linkage to the inositol, followed by three mannose residues. The first mannose is attached to the glucosamine by an α-1,4-linkage, the second mannose is attached to the first by an α-1,6-linkage, and the third mannose is attached to the second by an α-1,2-linkage. The phosphate of the phosphoethanolamine is attached to the 6 position of the third mannose, and the amino group of the ethanolamine is used in the attachment of the GPI anchor to the protein. Some species have some additional sugars that are added to this core structure, and the lipid moieties from the phosphatidylinositol also differ between species. In some cases, the GPI anchor is built on a sphingolipid instead of on a phospholipid. In *A. fumigatus*, GPI anchors have been shown to be attached to phosphoceramide (Fontaine et al., 2003). Within seconds after being released into the lumen of the ER, GPI-anchored proteins are recognized by the GPI anchor phosphoethanolamine transferase complex, which cleaves the protein at the omega site and transfers it onto the GPI anchor. A new covalent bond is formed between the amine group of the ethanolamine and the newly generated carboxyl group of the omega amino acid during the reaction.

Since over half of the cell wall proteins are GPI-anchored proteins and the GPI anchor plays a vital role in their movement through the secretory pathway, it is not surprising that mutations affecting the 20 or so genes encoding the enzymes and auxiliary factors needed for the biosynthesis of the GPI anchor are lethal in *S. cerevisiae*. Temperature-sensitive mutations affecting GPI anchor biosynthesis have been used to characterize the biosynthetic pathway in yeast (Eisenhaber et al., 2003; Kinoshita & Inoue, 2000).

In *N. crassa*, deletion mutants affecting some of the auxiliary factors are viable. These mutants have a tight colonial morphology, are susceptible to lysis in the absence of osmotic stabilization, and GPI-anchored cell wall proteins are missing from their cell walls (Bowman et al., 2006). Loss of GPI anchor biosynthesis in *S. pombe* has been shown to be lethal (Colussi & Orlean, 1997). In *C. albicans* and *A. fumigatus*, GPI anchor biosynthesis has been shown to be required for normal morphology and virulence (Li et al., 2007; Richard, 2002; Victoria, Kumar, & Komath, 2010). Clearly, the ability to synthesize the GPI anchor and attach the anchor to GPI-anchored proteins is critical for the formation of the fungal cell wall.

As previously mentioned in Section 3.4, the GPI anchor has been shown to be an important element in the cross-linking of cell wall proteins into the cell wall matrix in *S. cerevisiae* and *C. albicans*. The oligosaccharide portion of the GPI anchor has been shown to be attached to β-1,6-glucan (Kapteyn et al., 1996; Kollár et al., 1997; Lu et al., 1995). The question of how non-GPI-anchored cell wall proteins are incorporated into the cell wall matrix in *S. cerevisiae* and *C. albicans* remains to be elucidated.

3.7.2 N-linked and O-linked mannans and galactomannans

The majority of the cell wall mannans and galactomannans are found as posttranslational modifications of the cell wall proteins, and the posttranslational glycosylation of cell wall protein is essential for cell wall formation (Jin, 2012). However, lipid-linked galactomannan has been reported in *A. fumigatus* (Costachel et al., 2005; Latgé, 2009). High molecular weight mannans are found in *S. cerevisiae* and *C. albicans*, while galactomannans and xylomannans are found in the cell walls of the other characterized fungi (Table 2.2). Mannans and galactomannans occur as O-linked glycosylations and as additions to the N-linked oligosaccharides, and many of the enzymes that function in the synthesis of these oligosaccharides have been identified and characterized in *S. cerevisiae* (Despande, Wilkins, Packer, & Nevalainen, 2008).

In *S. cerevisiae* and *C. albicans*, synthesis of the O-linked mannans begins with a mannose residue being added to a serine or threonine residue in the amino acid sequence of the protein. The enzymes responsible for this addition, protein mannosyltransferases, have been identified in yeast, which has six such proteins (Pmt1p—Pmt6p) encoded in the genome. Additional mannoses are then added to the oligosaccharide by α-1,2-mannosyltransferases (Mnt1p, Ktr1p, Ktr3p) to create a short α-1,2-mannose chain. Finally, the O-linked oligosaccharides are terminated with α-1,3-mannoses, which are added by α-1,3-mannosyltransferases (Mnn1p, Mnt2p, and Mnt3p). In *S. pombe* and *N. crassa*, which have galactomannan as their O-linked oligosaccharide, homologs are found for the protein mannosyltransferases and the α-1,2-mannosyltransferases, but not for the side chain-generating α-1,3-mannosyltransferases. This is as expected, since the O-linked galactomannan structure contains an α-1,2-mannose chain and differs from the O-linked mannan by the presence of β-1,5-linked galactofuranose in the side chains instead of α-1,3-mannose side chains (Ballou, Ballou, & Ball, 1994; Nakajima, Yoshida, Nakamura, Hura, & Matsuda, 1984). *A. fumigatus* also has O-linked mannogalactans, and there is some evidence that the mannose chain has α-1,6-linkages instead of α-1,2-linkages (Leitão et al., 2003). However, *A fumigatus* does contain Mnt1, the enzyme to make the short α-1,2-mannan chain, and mnt-1 deletion mutants have cell wall defects and attenuated virulence (Wagener et al., 2008). The galactofuranosyltransferases responsible for adding the galactofuranose side chains have yet to be identified. Mutational analysis shows that mutations affecting the synthesis of O-linked oligosaccharides can have dramatic effects on the growth, morphology, and virulence of *C. neoformans*, *C. albicans*, *A. fumigatus*, and *N. crassa* (Buurman et al., 1998; Bowman, Piwowar, Ciocca, & Free, 2005; Olson, Fox, Wang, Alspaugh, & Buchanan, 2007; Wagener et al., 2008; Zhou et al., 2007). Since almost all cell wall proteins contain substantial amounts of O-linked oligosaccharides, one would suspect that a large number of proteins are affected when O-linked oligosaccharide synthesis is compromised, and many of these proteins may be degraded or rendered nonfunctional in the absence of O-linked oligosaccharide addition.

The cell wall proteins also have a substantial amount of N-linked glycosylation. Like all eukaryotes, fungi have all of the machinery to create the dolichol phosphate-attached *N*-acetylglucosamine 2:mannose 9:glucose 3 oligosaccharide and to transfer the oligosaccharide to aspargine (N) residues found in the context of N-X-S/T within the amino acid sequence of the protein (Stanley, Schachter, & Taniguchi, 2009). This addition occurs on the luminal face of the ER as the cell wall proteins are being cotranslationally extruded into the lumen of the ER. The removal of the three glucoses (which is tied to the protein folding process) and perhaps one of the mannoses is thought to occur in the fungi similarly to what has been shown in other eukaryotes (Cumming & Doering, 2009). The fungi do not generate the complex oligosaccharides found in vertebrates, and the N-linked GlcNAC2:mannose8 structure is not further processed. What occurs in *S. cerevisiae* and in *C. albicans* is the addition of a very large outer chain mannan to the N-linked oligosaccharide. This begins with the enzyme Och1p adding a single α-1,6-linked mannose to a specific mannose within the N-linked oligosaccharide (Cumming & Doering, 2009; Lehle, Eiden, Lehnert, Haselbeck, & Kopetzki, 1995; Nakanishi-Shindo, Nakayama, Tanaka, Toda, & Jigami, 1993). A long chain containing 50 or more α-1,6-mannoses is then built upon the initial mannose that was added by Och1p. Most of the enzymes needed for the elongation of this outer chain mannan (Van1p, Anp1p, Mnn9p, Mnn10p, and Mnn11p) are highly conserved across the fungi (Despande et al., 2008). In *S. cerevisiae* and *C. albicans*, the α-1,6-mannan chain is then further modified by the addition of side chains containing α-1,2-mannans which are terminated by α-1,3-mannoses. The final size of the structure can contain up to 200 mannose residues, and the mannans can account for up to 20% of the cell wall mass (Klis et al., 2006; Lesage & Bussey, 2006). The side chain α-1,2-mannoses are added by the Mnn2p and Mnn6p enzymes, which are conserved among the various fungal species. Finally, the terminal α-1,3-mannoses are added by Mnn1p, which is found only in *S. cerevisiae* and *C. albicans*.

Instead of the large outer chain mannan, *S. pombe* and *N. crassa* have N-linked galactomannan structures, which have not been as well characterized as the large outer chain mannan. The α-1,6-mannose chain for these galactomannans is much shorter than that of the yeast outer chain mannan, and the side chains contain α-1,2-mannose residues terminated by galactofuranoses (Ballou et al., 1994; Nakajima et al., 1984). The N-linked glucan from *C. neoformans* has also been recently analyzed and shown to be a xylomannan. It has a long α-1,6-mannan core with side chains consisting of α-1,2-mannans terminated with a xylose residue (Park et al., 2012). The *N. crassa* and *S. pombe* galactomannans, the *C. neoformans* xylomannan, and the *S. cerevisiae* and *C. albicans* outer chain mannans all share a similar structure with the terminal sugars being species-specific. The galactomannans isolated from *A. fumigatus* have been reported to have a somewhat different core structure (Fontaine et al., 2000). Instead of having a linear α-1,6-mannan core, the *A. fumigatus* core has been reported to have a repeating tetrameric mannan (α-1,2-mannose-α-1,6-mannose-α-1,2-mannose-α-1,2-mannose) with β-1,5-galactofuranoside side chains. The homologs of the characterized yeast enzymes for the creation of an α-1,6-mannose chain (Och1p, Van1p, Anp1p, and Mnn9p) and the enzymes for the α-1,2-mannose part of the side chain (Mnn2p and Mnn6p) are found throughout the fungi. As with the case for O-linked oligosaccharides, the enzymes that add the terminal galactose and xylose residues to the galactomannans and xylomannan remain to be identified, and there is no homolog for Mnn1p in *S. pombe, N. crassa, A. fumigatus,* or *C. neoformans* (Despande et al., 2008).

Recently, it has been shown that the N-linked galactomannan is required for the incorporation of integral cell wall proteins into the *N. crassa* cell wall matrix. Mutants lacking OCH-1, the enzyme that adds the initial mannose of the galactomannan to the N-linked oligosaccharide, are unable to incorporate GPI-anchored and non-GPI-anchored cell wall proteins into the cell wall (Maddi & Free, 2010). Similarly, *N. crassa* deletion mutant for MNN9 (mannan polymerase II, which is involved in building the α-1,6-mannose core) releases cell wall proteins into the growth medium (unpublished). The *och-1* and *mnn-9* mutants are dramatically affected in their growth morphologies. The *och-1* deletion mutant has a weak cell and is susceptible to lysis (Maddi & Free, 2010). Interestingly, *N. crassa* mutants that are lacking the α-1,6-mannanases, DFG5 and DCW1, pheno-copy the OCH-1 mutant, both in morphology and in being unable to incorporate cell wall proteins into the cell wall matrix (Maddi, Dettmann, Fu, Seiler, & Free, 2012; Maddi, Fu, & Free, 2012). Together, these findings suggest that the DFG5 and DCW1 mannanases function in cross-linking the *N. crassa* N-linked galactomannan into the cell wall matrix and incorporate cell wall proteins into the cell wall. This mechanism for incorporation of cell wall proteins may function in some other fungal species, but the available data suggests it is not used in *A. fumigatus.* The *A. fumigatus* Och1 mutant has a normal morphology and is not sensitive to cell wall stress (Kotz et al., 2010), which suggests that the N-linked oligosaccharide is not required for the incorporation of cell wall protein in *A. fumigatus.*

3.8. Functions ascribed to individual cell wall proteins
3.8.1 Cell wall biogenesis
A major fraction of the identified cell wall proteins are thought to function in cross-linking the various elements of the wall together to create the three-dimensional cell wall matrix and to remodel the cell wall to allow for branching and anastomosis. These enzymes are unique to the fungi, and the genes encoding them have been found to be conserved in fungal genomes. These enzymes function as glycosyl hydrolases (glucanases, man-nanases, etc.) to cleave cell wall polymers and as transglycosidases (glycosyltransferases) to create a new glycosidic bond between the cleaved polymer and another polymer in the cell wall (Cabib et al., 2007; Goldman, Sullivan, Zakula, & Capobianco, 1995; Mouyna et al., 2000). Almost all of these enzymes are classified as glycosyl hydrolases in the CAZy

database (Cantarel et al., 2009). In almost all cases, these enzymes are encoded in multigene families, and multiple members of these families are expressed in the vegetative cell wall. This arrangement provides the cell wall with a redundancy of cross-linking activity and helps insure that the fungus has a fully functional cell wall. This redundancy is observed in genetic analyses, which show that in most cases, loss of a single enzyme does not give rise to the dramatic cell wall phenotypes observed with the loss of one of the major cell wall polymers, with the loss of GPI-anchoring, or with the loss of O-linked and N-linked oligosaccharide synthesis. Generally speaking, mutations affecting a single cell wall protein have a much more subtle effect, such as an increased sensitivity to cell wall perturbation reactions or a decreased growth rate. In some cases, double or triple mutants, in which two or more enzymes with a redundancy of functions are lost, have a much more severe and obvious phenotype. Different members of a multigene family can be expressed in different cell types, and genetic analysis demonstrates that they can function in a cell-type-specific manner. Some cell wall enzymes may also function in "softening" the cell wall during hyphal branching and in the process of cell-to-cell fusion, where cell wall synthesis and degradation have to be carefully controlled to allow for branch formation or cell fusion while avoiding cell lysis. Examples of some better characterized families of cell wall cross-linking enzymes are discussed below.

3.8.2 The Gas1p/GEL1/Phr1p family of β-1,3-glucanases

These enzymes (glycosyl hydrolase family 72) are the best characterized of all the cell wall enzymes. The enzymes are GPI-anchored, and Gas1p/GEL1/Phr1p homologs have been found in a number of different species. Members of this family of β-1,3-glucanases have been identified as major cell wall proteins in all five of the ascomycete species shown in Table 2.1 and in *C. neoformans* (Eigenheer et al., 2007). These enzymes can cleave β-1,3-glucans and also create new β-1,3-bonds, making them capable of lengthening and shortening β-1,3-polymers (Hartland et al., 1996; Mouyna et al., 2000). In *C. albicans*, the enzymes have been shown to be required for the proper cross-linking between β-1,3-glucan and β-1,6-glucan (Fonzi, 1999). The crystal structure of *S. cerevisiae* Gas2p has been elucidated and provides important insights into how the enzyme functions (Hurtado-Guerrero et al., 2009). One face of the protein has an extended binding pocket which accommodates up to 10 β-1,3-linked glucose residues. The cleavage site is such that an enzyme-linked intermediate is produced in the enzyme's active site. The intermediate bond is protected from water hydrolysis while the glucan fragments are bound to the enzyme. In a transglycosylation step, the "enzyme-linked" fragment is transferred to a second β-1,3-glucan to produce a lengthened β-1,3-glucan.

Mutational analysis of gas, gel, and phr mutants shows that their activity is needed to generate a normal cell wall. Fungi have multiple GAS/GEL/PHR genes in their genomes. S. cerevisiae has five GAS genes (Ragni, Coluccio, et al., 2007; Ragni, Fontaine, Gissi, Latgé, & Popolo, 2007), A. fumigatus has seven GEL genes (Gastebois, Fontaine, Latgé, & Mouyna, 2010; Gastebois, Mouyna, et al., 2010), C. albicans has five PHR genes (Calderon et al., 2010), S. pombe has four gas genes (De Medina-Redondo et al., 2008), and N. crassa has five gel genes (Borkovich et al., 2004). Mutational analysis has shown that in some cases, multiple gas/gel enzymes are expressed and have overlapping specificity. In other cases, different gas/gel enzymes are expressed in different cell types. These enzymes have been extensively studied in A. fumigatus and S. cerevisiae. In A. fumigatus, the GEL1, GEL2, and GEL4 are expressed during vegetative growth, and GEL4 is essential (Mouyna et al., 2005; Gastebois, Fontaine, et al., 2010, Gastebois, Mouyna, et al., 2010). Deletions of GEL2 give rise to an altered cell wall and a reduction in virulence (Mouyna et al., 2005). In S. cerevisiae, Gas1p and Gas5p are expressed in vegetative hyphae, while Gas2p and Gas4p are expressed during sporulation (Rolli et al., 2011). Deletion of GAS1 results in a reduced level of cell wall glucan, an increased amount of β-1,3-glucan being released to the medium, and in swollen spherical cells (Ram et al., 1998). While the deletion of GAS2 or GAS4 individually does not affect sporulation, the GAS2/GAS4 double deletion mutant is severely affected in sporulation and the formation of the spore cell wall, which indicates that the two enzymes have redundant specificities (Ragni, Coluccio, et al., 2007, Ragni, Fontaine, et al., 2007). In C. albicans, Phr1p and Phr2p are expressed in response to pH conditions, with Phr2p being expressed at low pH and Phr1p being expressed at higher pH. In a pH-dependent manner, Phr1p and Phr2p are required for normal morphology and virulence (Fonzi, 1999). Phr1p has also been recently shown to be required for the adhesion to abiotic substrates and to epithelial cells (Calderon et al., 2010). In S. pombe, Gas4p is essential for ascospore cell wall maturation (De Medina-Redondo et al., 2008). These findings indicate that, in some cases, fungi produce two or more Gas1p/GEL1/Phr1p enzymes with redundant or partially overlapping activities. These findings also demonstrate that different Gas1p/GEL1/Phr1p enzymes are cell-type specific and are dedicated to cell wall biogenesis in particular cell types.

3.8.3 The Crh/Utr family of β-1,3-glucanases

These enzymes (glycosyl hydrolase family 16) have been identified as major cell wall proteins in proteomic analyses of S. *cerevisiae, C. albicans, A. fumigatus*, and *N. crassa* cell walls (Table 2.1). Mutational analysis shows that this class of transglycosidases function in cross-linking the chitin and β-1,6-glucan polymers in *S. cerevisiae* (Cabib, Blanco, Grau, Rodriguez-Peña, & Arroyo, 2007). In *C. albicans*, the Crh enzymes have been shown to be needed to create attachment between chitin and the cell wall β-1,3-glucan (Pardini et al., 2006). *S. cerevisiae* has two related enzymes, Crh1p and Crh2p, and deletion of both enzymes is required to create a severe mutant phenotype (Cabib et al., 2007). *C. albicans* has three enzymes, Crh11p, Crh12p, and Utr2, and deletion of all three enzymes gives rise to a defective cell wall (Pardini et al., 2006). These results demonstrate that the Crh enzymes play an important role in cell wall biogenesis. The Crh enzymes have overlapping activities and represent a case of redundancy in cell wall biosynthetic functions. Unlike the case of the gas/gel enzymes which can change the length of the cell wall glucan, the Crh enzymes clearly create cross-linking between cell wall components.

3.8.4 The Bgl2p/Scw4p/Scw10p/Scw11p family of β-1,3-glucanases

Members of this family of enzymes (glycosyl hydrolase family 17) have been found in the *S. cerevisiae, C. albicans, N. crassa*, and *A. fumigatus* cell walls (Table 2.1). The enzymatic activity and reaction kinetics have been studied for purified Bgl2p from *S. cerevisiae* and *C. albicans* (Goldman et al., 1995). These enzymes function to cleave a disaccharide from the end of a β-1,3-glucan polymer and form an activated intermediate. The activated intermediate is then transferred to the nonreducing end of a second β-1,3-glucan to create a "kinked" polymer. The *S. cerevisiae* Bgl2p mutant has a near normal morphology, but has an increased level of chitin in its cell wall (Kalebina et al., 2003). *S. cerevisiae* mutants in Scw4p and Scw10p are sensitive to cell wall destabilizing agents, and the Scw4p/Scw10p double mutant showed a synergistic effect, suggesting that these two enzymes function in cell wall biogenesis (Sestak, Hagen, Tanner, & Strahl, 2004). Bgl2p mutants of *C. albicans* have an altered cell wall and have an attenuated virulence (Sarthy et al., 1997).

In *A. fumigatus*, two Bgl2 family members, AfBGT1 and AfBGT2, have been characterized (Gastebois et al., 2009). AfBGT1, like the *S. cerevisiae* and *C. albicans* Bgl2p enzymes, cleaves a disaccharide from the reducing end of a donor β-1,3-glucan and transfers the glucan to the nonreducing end of an acceptor β-1,3-glucan to generate a "kinked" polymer. The AfBGT2 enzyme also cleaves a disaccharide from the reducing end of a donor β-1,3-glucan, but can transfer the donor β-1,3-glucan to an internal site on an acceptor β-1,3-glucan to create a branched β-1,3-glucan molecule with a β-1,6-branch site (Gastebois, Fontaine, et al., 2010, Gastebois, Mouyna, et al., 2010). Thus, the AfBGT2 enzyme has the capability of creating a cross-linked β-1,3-glucan matrix. Interestingly, an *A. fumigatus* AfBGT1/AfBGT2 double mutant does not display an obvious cell wall mutant phenotype, suggesting that these enzymes are not essential for the creation of the cell wall. It is possible that the *A. fumigatus* cell may have some additional enzymes with functional redundancy that can function to generate a cross-linked β-1,3-glucan cell wall matrix.

3.8.5 The DFG5/DCW1 family of α-1,6-mannanases
These GPI-anchored enzymes from glycosyl hydrolase family 76 are critical for cell wall biosynthesis. The author's laboratory recently showed that the *N. crassa dfg5/dcw1* double mutant is unable to incorporate cell wall proteins into the cell wall (Maddi, Dettmann, et al., 2012; Maddi, Fu, et al., 2012). Previously, Maddi and Free (2010) demonstrated that the N-linked galactomannan, with its α-1,6-mannose core, was required for the covalent incorporation of cell wall proteins into the cell wall matrix. Together, these findings suggest that the α-1,6-mannan chains associated with the glycoproteins are being recognized by the DFG5 and DCW1 mannanases and are being used in cross-linking reactions. Presumably, the DFG5 and DCW1 enzymes cleave the α-1,6-mannan and in the process of doing so create new glycosidic bonds between glycoprotein-associated galactomannans and cell wall glucans. Thus, in *N. crassa*, the cell wall proteins are being tied into the matrix in essentially the same way that the chitin and glucans are being cross-linked together, by glycosylhydrolases/glycosyltransferases that cross-link cell wall polymers together. Mutational analysis shows that DFG5 and DCW1 have overlapping specificities in *N. crassa*, *S. cerevisiae*, and *C. albicans*, and that they are required for cell wall biogenesis (Kitagaki, Wu, Shimoi, & Ito, 2002; Spreghini, Davis, Subaran, Kim, & Mitchell, 2003; Kitagaki, Ito, & Shimoi, 2004; Maddi, Dettmann, et al., 2012; Maddi, Fu, et al., 2012). Similar to the situations discussed above for the Gas1p/GEL1/Phr1p, Crh, and Bgl2/Scw4/Scw10/Scw11 families of enzymes, the DFG5 and DCW1 enzymes are both produced during vegetative growth and represent yet another case of the fungi producing two or more cell wall cross-linking enzymes with a redundancy in their activities.

3.8.6 Chitinases

Enzymes from glycosyl family 18 have chitinase activity and may be involved in cell wall remodeling (Adams, 2004). Mass spectrometry has identified chitinases as major cell wall proteins in *S. cerevisiae, C. albicans*, and *N. crassa* (Table 2.1). While *S. cerevisiae* and *C. albicans* have three and four chitinase genes, respectively (Adams, 2004; Dünkler et al., 2005), *A. fumigatus* has at least 11 chitinase genes, which makes genetic analyses difficult. However, specific chitinases have been shown to be required for septation and cell separation. In *S. cerevisiae*, deletion of Cts1 affects septation (Kuranda & Robbins, 1991). During septation, chitin is synthesized between the mother and daughter cells, and Cts1p functions in degrading this chitin and facilitating the separation of mother and daughter cells. In *C. albicans*, deletion of Cht3 gives rise to a similar cell separation phenotype (Dünkler et al., 2005). Genetic analyses of the chitinases in *A. fumigatus* have failed to identify a chitinase function (Jacques et al., 2003; Alcazar-Fuoli et al., 2011), but *A. fumigatus* has at least 11 chitinases encoded in its genome. These chitinase genes are differentially regulated during growth and development (Taib, Pinney, Westhead, McDowall, & Adams, 2005). Although there is some evidence that the chitinases can participate in transglycosylation reactions (Xia et al., 2001; Jacques et al., 2003), the evidence to date suggests that the chitinases are mainly functioning in the degradation of cell wall chitin. In filamentous fungi, the chitinases may be important for hyphal branching and cell-to-cell fusion, but the number of chitinases may represent a situation where the simultaneous expression of several chitinases provides redundancy of function.

3.8.7 Structural proteins

A common set of structural proteins have been noted in a number of characterized cell walls suggesting that these proteins have common and important functions. In general, these proteins show lower levels of homology across the species than the enzymes discussed above. Many of these proteins are rich in serine and threonine residues and are likely to be heavily glycosylated with O-linked glycosylation. Perhaps the most abundant cell wall protein is Ecm33p, a GPI-anchored cell wall protein that is found in virtually all of the better characterized cell walls (Table 2.1). *S. cerevisiae* mutants lacking Ecm33p have a weakened cell wall and altered cell wall architecture (Pardo et al., 2004). In *C. albicans*, Ecm33p is required for normal cell architecture and virulence (Martinez-Lopez, Monteoliva, Diez-Orejas, Nombela, & Gil, 2004; Martinez-Lopez, Park, Myers, Gil, & Filler,

2006). In both *S. cerevisiae* and *C. albicans*, loss of Ecm33p induces the CWI pathway (see below) (Pardo et al., 2004), which demonstrates that Ecm33p is an important cell wall component. Studies in *S. pombe* also show that loss of Ecm33p affects cellular morphology and activates the CWI pathway (Takada et al., 2010). In *A. fumigatus*, loss of Ecm33p results in a conidial separation defect, hypervirulence, and resistance to caspofungin (β-1,3-glucan synthase inhibitor) (Chabane et al., 2006; Romano et al., 2006). The increased virulence and caspofungin resistance phenotypes may be due to activation of the CWI-MAP kinase pathway. These results clearly demonstrate that Ecm33p is an important cell wall protein, and may be the principal structural cell wall protein.

In *S. cerevisiae* and *C. albicans*, Pir proteins (proteins with internal repeats) have been found as major cell wall proteins (Table 2.1). These proteins have a signal peptide sequence, a cleavage site for the Kex2p serine protease near their N-terminus, multiple repeats of a consensus sequence (Q[IV]XDGQ [IVP]Q), and a conserved four cysteine domain at their C-terminus (De Groot, Ram, & Klis, 2005). These proteins are attached to the cell wall β-1,3-glucan and are released from the cell wall by treatment with mild alkali. The nature of the linkage between the protein and the glucan is unknown, but could be through O-linked oligosaccharides or some other alkali-sensitive linkage (Lesage & Bussey, 2006). Ecker, Deutzmann, Lehle, Mrsa, and Tanner (2006) suggest that a novel alkali-labile glutamate–glucose linkage may be responsible for cross-linking the Pir proteins into the cell wall. The quadruple Pir1p/Pir2p/Pir3p/Pir4p *S. cerevisiae* mutant has larger cells and shows pronounced susceptibility to cell wall inhibitors, demonstrating that the Pir proteins are important for a normal cell wall (Mrsa & Tanner, 1999). The *S. cerevisiae* Pir proteins have been shown to provide resistance against osmotin, a plant antifungal compound, and are important for long-term survival of stationary phase cells (Yun et al., 1997; Teparic, Stuparevic, & Mrsa, 2004).

3.8.8 Adhesins

Fungal cell walls contain proteins that allow them to stick to the substratum. For pathogenic fungi, these adhesins are vital to the infection process, but even for saprophytic fungi, the ability to assess the environment and to adhere to a nutrient-rich substratum is an important function performed by cell wall proteins. As might be expected, a variety of adhesins have been identified from fungal cell walls, and it is likely that many more adhesins will be discovered in the future. Clearly these adhesins will be of importance to the process of biofilm formation and cell adhesion. In *S. cerevisiae*, a-agglutinin and α-agglutinin are expressed during mating and facilitate cell–cell interactions (Huang, Dougherty, & Erdman, 2009). The a-agglutinin consists of two subunits, Aga1p and Aga2p. Aga1p and Aga2p are expressed at high levels of mating-type a cells in response to mating pheromone. The carboxyl terminus of AGA2 is a high affinity ligand for α-agglutinin. The α-agglutinin, Sag1p, is expressed on mating-type α cells and has a globular N-terminal region that binds to Aga2p. The agglutinins function to facilitate cell–cell recognition and binding during mating. *S. cerevisiae* also has five FLO genes, which participate in cell adhesion activities (Verstrepen & Klis; 2006; Dranginis, Rauceo, Coronado, & Lipke, 2007). Four of these, Flo1p, Flo5p, Flo9p, and Flo10p, are lectin-type proteins that bind mannose or mannose and glucose, and facilitate cell aggregation or flocculation between yeast cells. Flocculation is an important activity in the fermentation industry, where yeast cells flocculate at the end of a fermentation, and either settle to the bottom of the fermentation vessel or float on the surface. In both cases, flocculation allows for the removal of the yeast from the fermentation product. Flo11p, is closely related to the other Flo proteins, but instead of recognizing mannose, it recognizes a variety of other substrates (Verstrepen & Klis, 2006; Goossens & Willaert, 2012). Flo11p is responsible for adhesion to agar and plastic surfaces, and is likely to facilitate binding to plant materials. Flo11p also plays an important role in biofilm formation.

Cell adhesion has been extensively researched in *C. albicans* (Sundstrom, 2002; Verstrepen & Klis, 2006; Dranginis et al., 2007). As might be expected for a fungus that lives in a commensal relationship with humans and can become a pathogen, *C. albicans* produces a number of cell wall adhesins which are thought to allow it to adhere to a variety of host tissues. Table 2.1 shows that these adhesins are major cell wall proteins in *C. albicans*. The *C. albicans* genome has an eight member family of the Als genes encoding related GPI-anchored adhesins, and multiple members of the family are expressed, making it difficult to decipher the functions of the individual Als proteins (Verstrepen & Klis, 2006). Als1p is highly expressed during vegetative growth. The expression of Als1p and Als5p in *S. cerevisiae* demonstrates that these adhesins bind to mammalian cell surfaces and that Als1p and Als5p can bind to a broad range of peptides (Nobbs, Vickerman, & Jenkinson, 2010). Deletion of Als3p has shown that it plays a critical role during biofilm formation and can mediate binding to host tissues (Zhao et al., 2006; Liu & Filler, 2011). However, Als3p is not required for virulence (Cleary et al., 2011). Als4p has been shown to bind to vascular endothelial cells (Zhao, Oh, Yeater, & Hoyer, 2005). The Als adhesins are thought to be the primary *C. albicans* adhesins.

Hwp1p is a very interesting *C. albicans* cell wall adhesin. It facilitates adhesion of the fungus to human skin and oral epithelia as well as playing a role biofilm formation (Nobile, Nett, Andes, & Mitchell, 2006). During the differentiation of human skin, keratinocyte cell surfaces are covalently cross-linked together by transglutaminases, which generate isopeptide bonds between glutamine residues and lysine residues on targeted small proline-rich (SPR) proteins. Hwp1p is a GPI-anchored cell wall protein with a region that mimics these SPR target proteins, and the human trans-glutaminases cross-link it to skin keratinocytes and oral epithelial cells, effectively cross-linking the fungus onto the host surface (Staab, Bradway, Fidel, & Sundstrom, 1999). In addition to the Als proteins and Hwp1p, a number of other proteins have been implicated as adhesion proteins, but much less is known about these proteins. What is clear is that adhesion to the host is a major cell wall-mediated activity and a number of cell wall proteins participate in adhesion. Interestingly one major difference between, *Candida glabrata*, a pathogen which is very closely related to *S. cerevisiae*, and *S. cerevisiae*, is that *C. glabrata* produces an extensive repertoire of adhesion proteins (Roetzer, Gabaldón, & Schüller, 2011).

A fumigatus, being a filamentous fungus, may have less of a need for adhe-sins to mediate "intraspecies social interactions," but adhesion to host tissues is thought to be an important step in its pathogenesis. *A fumigatus* conidia stick tightly to pulmonary epithelial cells and to extracellular matrix compo-nents like laminin and fibronectin, and efforts to identify *A. fumigatus* pro-teins and polysaccharides that mediate this binding have been made (Sheppard, 2011). No known adhesins have been identified among the major cell wall proteins found on vegetative hyphae grown in culture (Table 2.1), but growth in the host may be needed to induce the expression of adhesins in *A. fumigatus*. A laminin-binding protein, AfCalA, has been identified, but its role in pathogenicity has not been demonstrated (Upadhyay, Mahajan, Ramjee, Basir, & Madan, 2009). Since negatively charged carbohydrate can inhibit the binding of *A. fumigatus* to laminin and to basal lamina, negatively charged cell wall oligosaccharides may also play a role in adhesion (Sheppard, 2011).

Little has been done in characterizing the adhesins in *C. neoformans, N. crassa*, and *S. pombe* and no adhesins have been identified among the major cell wall proteins (Table 2.1). Recently, *C. neoformans* has been shown to have an adhesin, Cfl1, which mediates flocculation (cell–cell aggregation) and hyphal growth (Wang, Zhai, & Lin, 2012). However, very little is known about *C. neoformans* binding to host tissues, except that the binding to lung tissues may be mediated by lectin-type proteins (Merkel & Scofield, 1997). In *S. pombe*, the Map4 protein has been shown to be a cell wall adhesin which functions to facilitate cell–cell interactions during mating (Sharifmoghadam & Valdivieso, 2008). What is apparent is that adhesins play critical roles in the behavior and cell biology of the fungi. Clearly, we have a lot to learn about fungal cell adhesion and how it is regulated during biofilm formation and during the life cycles of the fungi. This is a research area that is still in its infancy, and it will be exciting to learn how fungal cell adhesins function to regulate the adhesion of cells to a variety of substrata.

3.8.9 Receptors for signal transduction pathways: CWI

Important interactions between the environment and the cell occur within the cell wall, and these interactions must lead to the activation of signal transduction pathways to allow the fungus to respond to environment change. Cell wall sensors respond to environment changes and are important cell wall components. Among the best characterized cell wall sensors are the *S. cerevisiae* Wsc1p and Mid2p family of sensors that respond to cell wall stress and activate the CWI-MAP kinase signal transduction pathway (Klis et al., 2006; Levin, 2011; Miermont, Uhlendorf, McClean, & Hersen, 2011). Although Wsc1p and Mid2p are transmembrane proteins, they are clearly associated with the cell wall where they function to assess the status of the cell wall (Levin, 2011; Miermont et al., 2011). Activation of the CWI signal transduction pathway increases the synthesis of the cell wall polymers in response to damage to the cell wall. The pathway also activates the synthesis of additional cell wall proteins to further strengthen the cell wall and prevent lysis. Similar MAP kinase signal transduction pathways operate in response to osmotic stress and pheromone signaling to control hyphal growth. In the osmotic stress pathway, the two transmembrane mucin proteins, Msb2p and Hkr1p, act as osmotic sensors and are associated with the Sho1p transmembrane protein (Miermont et al., 2011). Acting as a complex, these proteins activate a MAP kinase pathway that regulates gene expression in response to osmotic stress. The CWI and osmotic stress (high osmolarity glycerol) MAP kinase signal transduction pathways act as "safety systems" that insure that the cell maintains a fully functional cell wall. These signal transduction pathways highlight the importance of the cell wall for the survival, growth, and morphology of the fungal cells. The presence of the CWI pathway allows the cell to compensate for the loss of a cell wall protein (Levin, 2011), and "masks" the phenotype that might otherwise be observed with the loss of an individual cell wall protein. An interesting situation has been found to exist in *N. crassa*, where two sensors have been shown to activate the CWI pathway. One of these sensors, the transmembrane protein WSC-1, is thought to activate the pathway in response to cell wall damage, and the other sensor, a small GPI-anchored protein called HAM-7, is thought to activate the pathway during the process of anastomosis (cell-to-cell fusion) (Maddi, Dettmann, et al., 2012; Maddi, Fu, et al., 2012). Although most of these sensors might be considered transmembrane proteins, they clearly function as cell wall elements and are critical for the creation of a functional cell wall. Mutations in these sensors or in their MAP kinase pathways give rise to severe cell wall defects (Levin, 2011; Miermont et al., 2011).

3.8.10 Other cell wall proteins

An examination of the cell wall proteins identified by proteomics shows the presence of proteins involved in oxygen free-radical detoxification in both *C. albicans* and *N. crassa* (Table 2.1). Cell wall superoxide dismutases and catalases play an important role in allowing cells to deal with oxygen free-radicals in their environment. Free radicals are often employed as part of a host defense system, and the ability to detoxify free radicals could be important for pathogenic fungi.

Proteases were also identified by proteomic analysis in the *C. albicans* and *C. neoformans* cell walls (Eigenheer et al., 2007). In *C. albicans*, genetic analysis shows that the identified aspartyl protease, Sap1p, is closely related to the *S. cerevisiae* yapsins (Monod, Hube, Hess, & Sanglard, 1998). The *S. cerevisiae* yapsins have been genetically shown to be required for CWI (Krysan, Ting, Abeijon, Kroos, & Fuller, 2005). The Yps1p yapsin has been shown to function in the proteolytic cleavage of Msb2p, a receptor for the osmotic stress MAP kinase signal transduction pathway (Vadaie et al., 2008). A large number of proteases were identified in a proteomic analysis of the *C. neoformans* cell wall and secretome (Eigenheer et al., 2007). The production of these proteases by the pathogens may reflect the importance of protease activity for *C. albicans* and *C. neoformans* pathogenicity.

The proteomic analyses show that cell wall cross-linking enzymes, the structural proteins, the adhesins, the oxygen free-radical detoxification proteins, and the proteases were each found in more than one of fungal cell walls examined (Table 2.1) and are likely to mediate cell wall functions that are common to all fungal cell walls. In addition to these proteins, individual fungal cell walls contained iron acquisition proteins and a number of conserved hypothetic proteins (see references listed in Table 2.1). Generally speaking, these proteins are identified by fewer peptide fragments in the proteomic analyses and are likely to be found in lower concentrations than the major proteins listed in Table 2.1. In addition to the proteins identified by proteomics, genetic analyses have identified additional cross-linking type enzymes that are important for cell wall biogenesis (Gómez-Esquer et al., 2004; Norice, Smith, Solis, Filler, & Mitchell, 2007). As we learn more about the fungal cell wall, it is almost certain that specific activities will be ascribed to many of the conserved hypothetical cell wall proteins, and we will come to better appreciate the diversity of functions carried out by cell wall proteins.

4. WELL-CHARACTERIZED FUNGAL CELL WALLS

A great deal of work and a large literature exists for the S. cerevisiae cell wall and for the cell walls of the pathogens A. fumigatus and C. albicans. Somewhat smaller, but still helpful, literatures exist for the pathogen C. neoformans, and for S. pombe and N. crassa. Aspects of each of these cell walls were compared and contrasted when the individual cell wall components were discussed. A short overview of each of these cell walls will be given here. These short descriptions are helpful in assessing how the different cell wall components are used by the fungi, and how these components vary from species to species. Table 2.2 contains a summary of the composition of the cell walls from these well-characterized fungi.

4.1. The S. cerevisiae cell wall

The S. cerevisiae cell wall consists largely of β-1,3-glucan with lesser amounts of β-1,6-glucan. The wall also contains a small amount of chitin, which is found in the bud scar formed during the separation of mother and daughter cells during budding growth. The β-1,6-glucan has been shown to cross-link the β-1,3-glucans together to create a matrix (Klis et al., 2006; Lesage & Bussey, 2006). β-1,6-Glucan has also been shown to be associated with the glycoprotein released from the cell wall by laminarinase (β-1,3-glucanase) digestion indicating that the protein is tied into the cell via β-1,6-glucan. The cell wall proteins contain N-linked outer chain mannans with up to 200 mannose residues and smaller O-linked mannans. GPI-anchored proteins are incorporated into the cell wall through β-1,6-glucans that are cross-linked onto the oligosaccharides found in the GPI anchor (Lu et al., 1995; Kapteyn et al., 1996; Kollár et al., 1997). Melanin, α-1,3-glucan, and galactomannans are absent from the *S. cerevisiae* cell wall. The yeast wall is organized such that the outer surface of the wall contains large amounts of the outer chain mannan and glycoprotein, with the glucans and chitin being more concentrated in the portion of the cell wall adjacent to the plasma membrane.

4.2. The *C. albicans* cell wall

In most aspects, the *C. albicans* cell wall resembles the *S. cerevisiae* cell wall (Ruiz-Herrera et al., 2006; Chaffin, 2008). The major constituents are β-1,3-glucan and β-1,6-glucan, with a small amount of chitin being present. The β-1,6-glucan has been shown to cross-link the β-1,3-glucan polymers. During host infection, the *C. albicans* wall contains melanin, which is thought to help strengthen the wall. *C. albicans* makes N-linked outer chain mannans. The wall lacks α-1,3-glucan and galactomannans. O-linked mannans are found associated with the cell wall glycoproteins. Proteomic analysis of hyphal and yeast form cells shows that there are some protein differences between these two cell types (Maddi et al., 2009; Sosinska et al., 2011; Heilmann et al., 2011).

4.3. The *A. fumigatus* cell wall

There is a very large literature describing the *A. fumigatus* cell wall (Latgé et al., 2005; Latgé, 2007), which contains some unique features. The major elements in the *A. fumigatus* cell wall are β-1,3-glucans, a mixed β-1,3-/β-1,4-glucan, α-1,3-glucan, and chitin. The levels of chitin are much higher in *A. fumigatus* than in the other characterized cell walls, and the mixed β-1,3/1,4-glucan may be unique to the *A. fumigatus* wall. β-1,6-Glucans are absent from the cell wall. The cell wall proteins are modified by N-linked galactomannans and O-linked galactomannans. However, the structure of the core mannan portion of the *A. fumigatus* N-linked galactomannan may be different from the galactomannan structures found in other species. Instead of having an α-1,6-mannan core, the *A. fumigatus* galactomannans have been reported to have a repeating tetramannose core that contains both α-1,2- and α-1,6-mannose linkages (Fontaine et al., 2000; Bernard & Latgé, 2001). The *A fumigatus* cell wall also contains melanin as an essential component.

4.4. The *S. pombe* cell wall

The *S. pombe* cell wall contains β-1,3-glucan and α-1,3-glucans as its major elements. The vegetative cell wall lacks chitin, but chitin has been found as a component of the conidial cell wall. β-1,6-Glucan is not present in the wall. Melanin is also absent from the *S. pombe* cell wall. The *S. pombe* cell wall proteins have N-linked galactomannans and O-linked galactomannans associated with them.

4.5. The *N. crassa* cell wall

The *N. crassa* cell wall has β-1,3-glucan as its major component (Maddi et al., 2009). The wall contains 4% chitin (S.J. Free, unpublished data), which is much less than found in *A. fumigatus* but more than in the yeast species. Sugar linkage data suggest that the wall may also have some mixed β-1,3-/β-1,4-glucan. β-1,6-Glucan is absent from the *N. crassa* cell wall. α-1,3-Glucan is restricted to the conidial cell wall and not found in the vegetative cell wall. The *N. crassa* cell wall glycoproteins contain N-linked galactomannan and O-linked galactomannan. Melanin is found in some specialized cell types in *N. crassa*, but is not present in the vegetative hyphae.

4.6. The *C. neoformans* cell wall

What is most remarkable about the *C. neoformans* cell wall is the presence of the capsule, a large zone of carbohydrate polymer found exterior to the canonical cell wall. Although not discussed in this review, the capsule has been extensively characterized and its composition and structure have been elucidated (Doering, 2009). The presence of the capsule complicates the characterization of the cell wall. The cell wall does contain β-1,3-glucans, β-1,6-glucans, and α-1,3-glucans. The α-1,3-glucans have been shown to function as the attachment site for the capsule to the cell wall. The *C. neoformans* glycoproteins have N-linked xylomannans (Park et al., 2012). Like the other two pathogens, *C. neoformans* produces melanized cell walls during host infection.

5. CONCLUSIONS: VARIABILITY IN CELL WALL COMPONENTS AND REDUNDANCY IN CELL WALL BIOSYNTHESIS

Although only a handful of fungal cell walls have been characterized, it is clear that a great deal of variability can exist in the components used in constructing the cell wall. The presence and importance of the various cell wall components (β-1,3-glucan, mixed β-1,3/1,4-glucan, β-1,6-glucan, chitin, α-1,3-glucan, melanin, mannans, and galactomannans) have been examined, and the information from the different species is summarized in Table 2.2. As is apparent from Table 2.2 and the narratives for the different species, there are some interesting differences in the polymer composition of the different fungal cell walls. Although each of the walls examined consists of a cross-linked matrix of polysaccharide polymers, there is a great deal of variety in which polymers are present in a cell wall. A much greater degree of variability would almost certainly be uncovered if we had data not only for the vegetative cell walls but also from the cell walls of the other cell types generated during the life cycles of these fungi. The data shows that β-1,3-glucan is the major cell wall polymer in all of the fungal cell walls examined. It is the only polymer that is universally found in the cell walls of the six fungal species examined. Each of the other constituents, chitin, mixed β-1,3-/β-1,4-glucan, β-1,6-glucan, α-1,3-glucan, and melanin are found in some, but not all of these fungal cell walls. Similarly, the structures of the N-linked and O-linked oligosaccharides vary, with *S. cerevisiae* and *C. albicans* having high mannan structures, *S. pombe, A. fumigatus,* and *N. crassa* having galactomannans, and *C. neoformans* having xylomannan. Although *S. cerevisiae* and *C. albicans* have β-1,6-glucan cross-linking the β-1,3-glucan, chitin, and glycoproteins together to generate a wall matrix, *A. fumigatus, S. pombe,* and *N. crassa* lack β-1,6-glucan and must therefore utilize an alternate method of cross-linking the matrix together.

Given the variability in cell wall polymer composition, it may be a little surprising that all of the fungal species seem to contain a similar array of the cross-linking enzymes described in Section 3.7 and shown in Table 2.1. Each species produces a number of different putative cross-linking enzymes, and only those that are found in more than one of the species are shown in Table 2.1. As described in the preceding sections, these enzymes are able to create a number of different types of cross-links between the cell wall glucans, chitin, and glycoproteins. As shown for the Gas1p/GEL1/Phr1p,

Crh/Utr, Bgl2p/Scw4p/Scw10p/Scw11p, and DFG5/DCW1 families of cross-linking enzymes, the fungi can produce multiple closely-related cross-linking enzymes to provide for a redundancy of cell wall cross-linking activity. It would be interesting to know whether the enzymes from the different species have slightly different substrate specificities to allow them to effectively cross-link their species-specific array of cell wall polymers. A large body of genetic data clearly demonstrates the importance of the cross-linking enzymes in generating a functional three-dimensional cell wall matrix, but there are likely to be interesting differences between different fungi in the details of how this is accomplished. This is most easily illustrated by considering how glycoproteins are incorporated into the cell wall. There have been a number of different mechanisms identified for the incorporation of glycoprotein into the wall. One mechanism, identified in *S. cerevisiae*, is through an attachment of β-1,6-glucan to the oligosaccharide portion of the GPI anchor structure (Lu et al., 1995; Kapteyn et al., 1996; Kollár et al., 1997). The *S. cerevisiae* Pir proteins, which are not GPI-anchored, are attached to the β-1,3-glucan through an alkali-sensitive bond, and one possibility is that the attachment might be through the O-linked oligosaccharides (Lesage & Bussey, 2006; Ruiz-Herrera et al., 2006). In *N. crassa*, the DFG5 and DCW1 α-1,6-mannanases have been shown to cross-link the N-linked mannans and galactomannans present on GPI-anchored and non-anchored proteins into the cell wall matrix (Maddi, Dettmann, et al., 2012; Maddi, Fu, et al., 2012). There is compelling evidence that none of these glycoprotein cross-linking mechanisms are universally used by all of the examined fungi. The lack of β-1,6-glucan in many fungi precludes the use of β-1,6-glucan as a universal cross-linker. Mutation in the O-linked oligosaccharide pathway does not affect the incorporation of glycoprotein into the *N. crassa* cell wall (personal observation), indicating that O-linked oligosaccharides are not needed for protein incorporation into the *N. crassa* cell wall. Similarly, mutations affecting N-linked galactomannan pathway does not affect the cell wall of *A. fumigatus* (Kotz et al., 2010), indicating that *A. fumigatus* does not use N-linked oligosaccharides in incorporating its glycoproteins into the wall. The available evidence shows that there must be multiple ways that fungi can use to cross-link glycoproteins into their cell walls. What most of the proposed mechanisms for incorporation of protein into the wall have in common is that the cross-linking occurs between carbohydrate polymers (between glucans and GPI anchor, N-linked, or O-linked oligosaccharides) and almost certainly involves transglycosidase activity. Thus, cross-linking glycoproteins into the cell wall is a variation on the theme of cross-linking the glucan and chitin polymers together by transglycosidases.

Both proteomic and genetic analyses have shown that fungi produce multiple members of the glucan and chitin cross-linking enzyme families as major cell wall proteins. By producing multiple cross-linking enzymes with overlapping specificities, and by having a number of different ways in which the cell wall elements can be cross-linked together, the fungi insure the presence of a functional cell wall. The redundancy in cell wall biogenesis enzymes has clearly been selected over evolutionary time and is likely to contribute to the ability of the fungi to survive in a changing environment.

ACKNOWLEDGMENTS

I would like to thank Dr. Abhiram Maddi for critical reading of the chapter and James Stamos for help in chapter preparation. Funding for the S. J. F. laboratory has been provided by a grant from the UB Foundation and by grants R01GM0785879 and R03AI103897 from the National Institutes of Health.

REFERENCES

Adams, D. J. (2004). Fungal cell wall chitinases and glucanases. *Microbiology*, *150*, 2029–2035.

Aimanianda, V., Clavaud, C., Simenel, C., Fontaine, T., Delepierre, M., & Latgé, J.-P. (2009). Cell wall β-(1,6)-glucan of *Saccharomyces cerevisiae*. *The Journal of Biological Chemistry*, *284*, 13401–13412.

Alcazar-Fuoli, L., Clavaud, D., Lamarre, C., Aimanianda, V., Seidl-Seiboth, V., Mellado, E., et al. (2011). Functional analysis of the fungal/plant class chitinase family in *Aspergillus fumigatus*. *Fungal Genetics and Biology*, *48*, 418–429.

Arellano, M., Cartagena-Lirola, H., Hajibagheri, M. A. N., Durán, A., & Valdivieso, M. H. (2000). Proper ascospore maturation requires the chs1+ chitin synthase gene in *Schizosaccharomyces pombe*. *Molecular Microbiology*, *35*, 79–89.

Baker, L. G., Specht, C. A., & Lodge, J. K. (2011). Cell wall chitosan is necessary for virulence in the opportunistic pathogen *Cryptococcus neoformans*. *Eukaryotic Cell*, *10*, 1264–1268.

Ballou, C. E., Ballou, L., & Ball, G. (1994). *Schizosaccharomyces pombe* glycosylation mutant with altered cell surface properties. *Proceedings of the National Academy of Sciences of the United States of America*, *91*, 9327–9331.

Banks, I. R., Specht, C. A., Donlin, M. J., Gerik, K. J., Levitz, S. M., & Lodge, J. K. (2005). A chitin synthase and its regulator protein are critical for chitosan production and growth of the fungal pathogen *Crytococcus neoformans*. *Eukaryotic Cell*, *4*, 1902–1912.

Beauvais, A., Bruneau, J. M., Mol, P. C., Buitrago, M. J., Legrand, R., & Latgé, J. P. (2001). Glucan synthase complex of *Aspergillus fumigatus*. *Journal of Bacteriology*, *183*, 2273–2279.

Bernard, M., & Latgé, J.-P. (2001). *Aspergillus fumigatus* cell wall: Composition and biosynthesis. *Medical Mycology*, *39*, S9–S17.

Beth-din, A., & Yarden, O. (2000). The *Neurospora crassa* chs3 gene encodes an essential class I chitin synthase. *Mycologia*, *92*, 65–73.

Birkaya, B., Maddi, A., Joshi, J., Free, S. J., & Cullen, P. J. (2009). Role of the cell wall Integrity and filamentous growth mitogen-activated protein kinase pathways in cell wall remodeling during filamentous growth. *Eukaryotic Cell*, *8*, 119–133.

Bohn, J. A., & BeMiller, J. N. (1995). (1→3)-β-D-Glucans as biological response modifiers: A review of the structure-functional activity relationship. *Carbohydrate Polymers*, *28*, 3–14.

Boone, C., Sommer, S. S., Hensel, A., & Bussey, H. (1990). Yeast KRE genes provide evidence for a pathway of cell wall beta-glucan assembly. *The Journal of Cell Biology*, *110*, 1833–1843.

Borkovich, K. A., Alex, L., Yarden, O., Freitag, M., Turner, G., Read, N., et al. (2004). Lessons from the genome sequence of *Neurospora crassa*: Tracing the path from genomic blueprint to multicellular organization. *Microbiology and Molecular Biology*, *63*, 1–108.

Bowman, S. M., & Free, S. J. (2006). The structure and synthesis of the fungal cell wall. *Bioessays*, *28*, 799–808.

Bowman, S. M., Piwowar, A., Al Dabbous, M., Vierula, J., & Free, S. J. (2006). Mutational analysis of the glycosylphosphatidylinositol (GPI) anchor pathway demonstrates that GPI-anchored proteins are required for cell wall biogenesis and normal hyphal growth in *Neurospora crassa*. *Eukaryotic Cell*, *5*, 587–600.

Bowman, S. M., Piwowar, A., Ciocca, M., & Free, S. J. (2005). Mannosyltransferase is required for cell wall biosynthesis, morphology and control of asexual development in *Neurospora crassa*. *Mycologia*, *97*, 872–879.

Bruneau, J.-M., Magnin, T., Tagat, E., Legrand, R., Bernard, M., Diaquin, M., et al. (2001). Proteome analysis *Aspergillus fumigatus* identifies glycosylphosphotidylinositol-anchored proteins associated to the cell wall biosynthesis. *Electrophoresis*, *22*, 2812–2823.

Buurman, E. T., Westwater, C., Hube, B., Brown, A. J. P., Odds, F. C., & Gow, N. A. R. (1998). Molecular analysis of CaMnt1p, a mannosyltransferase important for adhesion and virulence of *Candida albicans*. *Proceedings of the National Academy of Sciences of the United States of America*, *95*, 7670–7675.

Cabib, E., Blanco, N., Grau, C., Rodriguez-Peña, J. M., & Arroyo, J. (2007). Crh1p and Crh2p are required for the cross-linking of chitin to β(1,6)glucan in the *Saccharomyces cerevisiae* cell wall. *Molecular Microbiology*, *63*, 921–935.

Cabib, E., Roh, D.-H., Schmidt, M., Crotti, L. B., & Varma, A. (2001). The yeast cell wall and septum as paradigms of cell growth and morphogenesis. *The Journal of Biological Chemistry*, *276*, 19679–19682.

Calderon, J., Zavrel, M., Ragni, E., Fonzi, W. A., Rupp, S., & Popolo, L. (2010). PHR1, a pH regulated gene of *Candida albicans* encoding a glucan-remodelling enzyme, is required for adhesion and invasion. *Microbiology*, *156*, 2484–2494.

Cantarel, B. L., Coutinho, P. M., Rancurel, C., Bernard, T., Lombard, V., & Henrissat, B. (2009). The Carbohydrate-Active- enzymes database (CAZy): An expert resource for glycogenomics. *Nucleic Acids Research*, *37*, D233–D238.

Cappellaro, C., Mrsa, V., & Tanner, W. J. (1998). New potential cell wall glucanases of *Saccharomyces cerevisiae* and their involvement in mating. *Journal of Bacteriology*, *180*, 5030–5037.

Castillo, L., Calvo, E., Martínez, A. I., Ruiz-Herrera, J., Valentin, E., Lopez, J. A., et al. (2008). A study of the *Candida albicans* cell wall proteome. *Proteomics*, *8*, 3871–3881.

Chabane, S., Sarfati, J., Ibrahim-Granet, O., Du, C., Schmidt, C., Mouyna, I., et al. (2006). Glycosylphosphatidylinositol-anchored Ecm33p influences conidial cell wall biogenesis in *Aspergillus fumigatus*. *Applied and Environmental Microbiology*, *72*, 3259–3267.

Chaffin, W. L. (2008). *Candida albicans* cell wall proteins. *Microbiology and Molecular Biology Reviews*, *72*, 495–544.

Chen, S. C.-A., Slavin, M. A., & Sorrell, T. C. (2011). Echinocandin antifungal drugs in fungal infections. *Drugs*, *71*, 11–41.

Christodoulidou, A., Bouriotis, V., & Thireos, G. (1996). Two sporulation-specific chitin deacetylase-encoding genes are required for the ascospore wall rigidity of *Saccharomyces cerevisiae*. *The Journal of Biological Chemistry*, *271*, 31420–31425.

Cleary, I. A., Reinhard, S. M., Miller, C. L., Murdoch, C., Thornhill, M. H., Lazzell, A. L., et al. (2011). *Candida albicans* ahesin Als3p is dispensible for virulence in the mouse model of disseminated candidiasis. *Microbiology, 157,* 1806–1815.

Colussi, P. A., & Orlean, P. (1997). The essential *Schizosaccharomyces pombe* gpi11 + gene complements bakers' yeast GPI anchoring mutant and is required for efficient cell separation. *Yeast, 13,* 139–150.

Cortés, J. C. G., Ishiguro, J., Durán, A., & Ribas, C. (2002). Localization of the (1,3)β-D-glucan synthase catalytic subunit homologue Bgs1p/Cps1p from fission yeast suggests that it is involved in septation, polarized growth, mating, spore wall formation and spore germination. *Journal of Cell Science, 115,* 4081–4096.

Costachel, C., Coddeville, B., Latgé, J.-P., & Fontaine, T. (2005). Glycosylphosphatidyl-inositol-anchored fungal polysaccharide in *Aspergillus fumigatus*. *The Journal of Biological Chemistry, 280,* 39835–39842.

Cumming, R. D., & Doering, T. L. (2009). Chapter 21. Fungi. In A. Varki, R. D. Cummings, J. D. Esko, H. H. Freeze, P. Stanley, C. R. Bertozzi, G. W. Hart & M. E. Etzler (Eds.), *Essentials of glycobiology* (pp. 309–320). Cold Spring Harbor, NY: Cold Spring Harbor Press.

De Groot, P. W. J., Brandt, B. W., Horiuchi, H., Ram, A. F. J., de Koster, C. G., & Klis, F. M. (2009). Comprehensive genomic analysis of cell wall genes in *Aspergillus nidulans*. *Fungal Genetics and Biology, 46,* S72–S81.

De Groot, P. W. J., De Boer, A. D., Cunningham, J., Dekker, H. L., de Jong, L., Hellingwerf, K. J., et al. (2004). Proteomic analysis of *Canida albicans* cell wall reveals covalently bound carbohydrate-active enzymes and adhesins. *Eukaryotic Cell, 3,* 955–965.

De Groot, P. W. J., Hellingwerf, K. J., & Klis, F. M. (2003). Genome-wide identification of fungal GPI proteins. *Yeast, 20,* 781–796.

De Groot, P. W. J., Ram, A. R., & Klis, F. M. (2005). Features and functions of covalently linked proteins in fungal cell walls. *Fungal Genetics and Biology, 42,* 657–675.

De Groot, P. W. J., Yin, Q. Y., Weig, M., Sosinska, G. S., Klis, F. M., & De Koster, C. G. (2007). Mass Spectrometric identification of covalently bound cell wall proteins from the fission yeast *Schizosaccharomyces pombe*. *Yeast, 24,* 267–278.

De Medina-Redondo, M., Arnáiz-Pita, Y., Fontaine, T., Del Rey, F., Latgé, J. P., & De Aldana, C. R. (2008). The β-1,3-glucanosyltransferase gas4p is essential for ascospore wall maturation and spore viability in *Schizosaccharomyces pombe*. *Molecular Microbiology, 68,* 1283–1299.

Despande, N., Wilkins, M. R., Packer, N., & Nevalainen, H. (2008). Protein glycosylation pathways in filamentous fungi. *Glycobiology, 18,* 626–637.

Doering, T. L. (2009). How sweet it is! Cell wall biogenesis and polysaccharide capsule formation in *Cryptococcus neoformans*. *Annual Review of Microbiology, 63,* 223–247.

Douglas, C. M., Foor, F., Marriman, J. A., Morin, N., Nielsen, J. B., Dahl, A. M., et al. (1994). The *Saccharomyces cerevisiae* FKS1 (ETG1) gene encodes an integral membrane protein which is a subunit of 1,3-β-D-glucan synthase. *Proceedings of the National Academy of Sciences of the United States of America, 91,* 12907–12911.

Dranginis, A. M., Rauceo, J. M., Coronado, J. E., & Lipke, P. N. (2007). A biochemical guide to yeast adhesins: Glycoproteins for social and antisocial occasions. *Microbiology and Molecular Biology Reviews, 71,* 282–294.

Dünkler, A., Walther, A., Specht, C. A., & Wendland, J. (2005). *Candida albicans* CHT3 encodes the functional homolog of the Cts1 chitinase of *Saccharomyces cerevisiae*. *Fungal Genetics and Biology, 42,* 935–947.

Ebanks, R. O., Chisholm, K., McKinnon, S., Whiteway, M., & Pinto, D. M. (2006). Proteomic analysis of *Candida albicans* yeast and hyphal cell wall and associated proteins. *Proteomics, 6,* 2147–2156.

Ecker, M., Deutzmann, R., Lehle, L., Mrsa, V., & Tanner, W. (2006). Pir proteins of *Saccharomyces cerevisiae* are attached to β-1,3-glucan by a new protein-carbohydrate linkage. *The Journal of Biological Chemistry, 281*, 11523–11529.

Eigenheer, R. A., Lee, Y. L., Blumwald, E., Phinney, B. S., & Gelli, A. (2007). Extracellular glycosylphosphatidylinositol-anchored mannoproteins and proteases of *Cryptococcus neoformans*. *FEMS Yeast Research, 7*, 499–510.

Eisenhaber, B., Maurer-Stroh, S., Novatchkova, M., Schneider, G., & Eisenhaber, F. (2003). Enzymes and auxiliary factors for GPI lipid anchor biosynthesis and post-translational transfer to proteins. *Bioessays, 25*, 367–385.

Eisenhaber, B., Schneider, G., Wildpaner, M., & Eisenhaber, F. (2004). A sensitive predictor for potential GPI lipid modification sites in fungal protein sequences and its application for genome-wide studies for *Aspergillus nidulans, Candida albicans, Neurospora crassa, Saccharomyces cerevisiae*, and *Schizosaccharomyces pombe*. *Journal of Molecular Biology, 337*, 243–253.

Eisenman, H. C., & Casadevall, A. (2012). Synthesis and assembly of fungal melanin. *Applied Microbiology and Biotechnology, 93*, 931–940.

Ferguson, M. A. J., Kinoshita, T., & Hart, G. W. (2009). Chapter 11: Glycosylphosphatidylinositol anchors. In A. Varki, R. D. Cummings, J. D. Esko, H. H. Freeze, P. Stanley, C. R. Bertozzi, G. W. Hart & M. E. Etzler (Eds.), *Essentials of glycobiology* (pp. 143–161). Cold Spring Harbor, NY: Cold Spring Harbor Press.

Fontaine, T., Beauvais, A., Loussert, C., Thevenard, B., Fulgsang, C., Ohno, N., et al. (2010). Cell wall α1-3glucans induce the aggregation of germinating conidia of *Aspergillus fumigatus*. *Fungal Genetics and Biology, 47*, 707–712.

Fontaine, T., Magnin, T., Melhert, A., Lamont, D., Latgé, J.-P., & Ferguson, M. A. J. (2003). Structures of the glycosylphosphatidylinositol membrane anchors from *Aspergillus fumigatus* membrane proteins. *Glycobiology, 13*, 169–177.

Fontaine, T., Simenel, C., Dubreucq, G., Adam, O., Delepierre, M., Lemoine, J., et al. (2000). Molecular organization of the alkali-insoluble fraction of the *Aspergillus fumigatus* cell wall. *The Journal of Biological Chemistry, 275*, 27594–27607.

Fonzi, W. A. (1999). *PHR1* and *PHR2* of *Candida albicans* encode putative glycosidases required for proper cross-linking of β-1,3-glucan and β-1,6-glucan. *Journal of Bacteriology, 181*, 7070–7079.

Frost, D. J., Brandt, K., Capobianco, J., & Goldman, R. (1994). Characterization of (1,3)-β glucan synthase in *Candida albicans*: Microsomal assay from the yeast or mycelia morphological forms and a permeabilized whole-cell assay. *Microbiology, 140*, 2239–2246.

Fuentes, A. M., Connerton, I., & Free, S. J. (1994). Production of tyrosinase defective mutants of *Neurospora crassa*. *Fungal Genetics Newsletter, 41*, 38–39.

Gastebois, A., Clavaud, D., Aimanianda, V., & Latgé, J.-P. (2009). *Aspergillus fumigatus*: Cell wall polysaccharides, their biosynthesis and organization. *Future Microbiology, 4*, 583–595.

Gastebois, A., Fontaine, T., Latgé, J.-P., & Mouyna, I. (2010). β(1,3)Glucosyltransferase Gel4p is essential for *Aspergillus fumigatus*. *Eukaryotic Cell, 9*, 1294–1298.

Gastebois, A., Mouyna, I., Simenel, C., Clavaud, C., Coddeville, B., Delepierre, M., et al. (2010). Characterization of a new β(1,3)-Glucan branching activity of *Aspergillus fumigatus*. *The Journal of Biological Chemistry, 285*, 2386–2396.

Gilbert, N. M., Donlin, M. J., Gerik, K. J., Specht, C. A., Djordjevic, J. T., Wilson, C. F., et al. (2010). KRE genes are required for β-1,6-glucan synthesis, maintenance of capsule architecture and cell wall protein anchoring in *Cryptococcus neoformans*. *Molecular Microbiology, 76*, 517–534.

Goldman, R. C., Sullivan, P. A., Zakula, D., & Capobianco, J. O. (1995). Kinetics of β-1,3 glucan interaction at the donor and acceptor sites of the fungal glucosyltransferase encoded by the *BGL2* gene. *European Journal of Biochemistry, 227*, 372–378.

Gómez-Esquer, F., Rodríguez, J. M., Díaz, G., Rodriguez, E., Briza, P., Nombela, C., et al. (2004). CRR1, a gene encoding a putative transglycosidase, is required for proper spore wall assembly in *Saccharomyces cerevisiae*. *Microbiology*, *150*, 3269–3280.

Goossens, K. V. Y., & Willaert, R. G. (2012). The N-terminal domain of the Flo11 protein from *Saccharomyces cerevisiae* is an adhesion without mannose-binding activity. *FEMS Yeast Research*, *12*, 78–87.

Grün, C. J., Hochstenbach, F., Humbel, B. M., Veerkleij, A. J., Sietsma, J. H., Klis, F. M., et al. (2004). The structure of cell wall α-glucan from fission yeast. *Glycobiology*, *15*, 245–257.

Hartland, R. P., Fontaine, T., Debeaupuist, J.-P., Simenel, C., Delepierre, M., & Latgé, J.-P. (1996). A novel β-(1,3)-glucanosyltransferase from the cell wall of *Aspergillus fumigatus*. *The Journal of Biological Chemistry*, *271*, 26843–26849.

Heilmann, C. J., Sorgo, A. G., Siliakus, A. R., Dekker, H. L., Brul, S., de Koster, C. G., et al. (2011). Hyphal induction in the human fungal pathogen *Candida albicans* reveals a characteristic wall protein profile. *Microbiology*, *157*, 2297–2307.

Hochstenbach, F., Klis, F. M., Van Den Ende, H., Van Donselaar, E., Peters, P. J., & Klausner, R. D. (1998). Identification of a putative alpha-glucan synthase essential for cell wall construction and morphogenesis in fission yeast. *Proceedings of the National Academy of Sciences*, *95*, 9161–9166.

Horiuchi, H. (2009). Functional diversity of chitin syntases of *Aspergillus nidulans* in hyphal growth, conidiophores development and septum formation. *Medical Mycology*, *47*, S47–S52.

Huang, G., Dougherty, S. D., & Erdman, S. E. (2009). Conserved WCPL and CX4C domains mediate several mating adhesion interactions in *Saccharomyces cerevisiae*. *Genetics*, *182*, 173–189.

Hurtado-Guerrero, R., Schüttelkopf, A. W., Mouya, I., Ibrahim, A. F. M., Shepherd, S., Fontaine, T., et al. (2009). Molecular mechanisms of yeast cell wall glucan remodeling. *The Journal of Biological Chemistry*, *284*, 8461–8469.

Insenser, M. R., Hernáez, M. L., Nombela, C., Molina, M., Molero, G., & Gil, C. (2010). Gel and gel-free proteomics to identify *Saccharomyces cerevisiae* cell surface proteins. *Journal of Proteomics*, *73*, 1183–1195.

Jacobson, E. S. (2000). Pathogenic roles for fungal melanins. *Clinical Microbiology Reviews*, *13*, 708–717.

Jacques, A. K., Fukamizo, T., Hall, D., Barton, R. C., Escott, G. M., Parkinson, T., et al. (2003). Disruption of the gene encoding the ChiB1 chitinase of *Aspergillus fumigatus* and characterization of the recombinant gene product. *Microbiology*, *149*, 2931–2939.

James, P. G., Chemiak, R., Jones, R. G., Stortz, C. A., & Reiss, E. (1990). Cell-wall glucans of *Cryptococcus neoformans* Cap67. *Carbohydrate Research*, *198*, 23–28.

Jin, C. (2012). Protein glycosylation in *Aspergillus fumigatus* is essential for cell wall synthesis and serves as a promising model for multicellular eukaryotic development. *International Journal of Microbiology*, *2012*, article ID 654251.

Kalebina, T. S., Farkas, V., Laurinavichiute, D. K., Gorlovoy, P. M., Fominov, G. V., Bartek, P., et al. (2003). Deletion of BGL2 results in an increased chitin level in the cell wall of *Saccharomyces cerevisiae*. *Antonie Van Leeuwenhoek*, *84*, 179–184.

Kapteyn, J. C., Montijn, R. C., Vink, E., de la Cruz, J., Llobell, A., Douwes, J. E., et al. (1996). Retention of *Saccharomyces cerevisiae* cell wall proteins through a phosphodiester-linked β-1,3-/β-1,6-glucan heteropolymer. *Glycobiology*, *6*, 337–345.

Karkowska-Kuleta, J., Rapala-Kozik, M., & Kozik, A. (2009). Fungi pathogenic to humans: Molecular bases of virulence of *Candida albicans*, *Cryptococcus neoformans* and *Aspergillus fumigatus*. *Acta Biochimica Polonica*, *56*, 211–224.

Kinoshita, T., & Inoue, N. (2000). Dissecting and manipulating the pathway for glycosylphosphatidylinositol-anchor biosynthesis. *Current Opinion in Chemical Biology*, *4*, 632–638.

Kitagaki, H., Ito, K., & Shimoi, H. (2004). A temperature-sensitive *dcw1* mutant of *Saccharomyces cerevisiae* is cell cycle arrested with small buds which have aberrant cell walls. *Eukaryotic Cell*, *3*, 1297–1306.

Kitagaki, H., Wu, H., Shimoi, H., & Ito, K. (2002). Two homologous genes, DCW1 (YKL046c) and DFG5, are essential for cell growth and encode glycosylphosphatidylinositol (GPI)-anchored membrane proteins required for cell wall biogenesis in *Saccharomyces cerevisiae*. *Molecular Microbiology*, *46*, 1011–1022.

Klis, F. M., Boorsma, A., & de Groot, P. W. J. (2006). Cell wall construction in *Saccharomyces cerevisiae*. *Yeast*, *23*, 185–202.

Klis, F. M., Sosinska, G. J., de Groot, P. W. J., & Brul, S. (2009). Covalently linked cell wall proteins of *Candida albicans* and their role in fitness and virulence. *FEMS Yeast Research*, *9*, 1013–1028.

Klutts, J. S., Yoneda, A., Reilly, M. C., Bose, I., & Doering, T. (2006). Glycosyltransferases and their products: Cryptococcal variations on fungal themes. *FEMS Yeast Research*, *6*, 499–512.

Kollár, R., Petráková, E., Ashwell, G., Robbins, P. W., & Cabib, E. (1995). Architecture of the yeast cell wall: The linkage between chitin and β(1,3)-glucan. *The Journal of Biological Chemistry*, *270*, 1170–1178.

Kollár, R., Reinhold, B. B., Petráková, E., Yeh, H. J. C., Ashwell, G., Drgonová, J., et al. (1997). Architecture of the yeast cell wall: β(1,6)-glucan interconnects mannoprotein, β(1,3)-glucan, and chitin. *The Journal of Biological Chemistry*, *272*, 17762–17775.

Kotz, A., Wagener, J., Engel, J., Routier, F. H., Echtenacher, B., Jacobsen, I., et al. (2010). Approaching the secrets of N-glycosylation in *Aspergillus fumigatus*: Characterization of the AfOch1 protein. *PLoS One*, *5*, e15729.

Krysan, D. J., Ting, E. L., Abeijon, C., Kroos, L., & Fuller, R. S. (2005). Yapsins are a family of aspartyl proteases required for cell wall integrity in *Saccharomyces cerevisiae*. *Eukaryotic Cell*, *4*, 1364–1374.

Kuranda, M. J., & Robbins, P. W. (1991). Chitinase is required for cell separation during growth of *Saccharomyces cerevisiae*. *The Journal of Biological Chemistry*, *266*, 19758–19767.

Langfelder, K., Jahn, B., Gehringer, H., Schmidt, A., Wanner, G., & Brakhage, A. A. (1998). Identification of a polyketide synthase gene (pksP) of *Aspergillus fumigatus* involved in conidial pigment biosynthesis and virulence. *Medical Microbiology and Immunology*, *187*, 79–89.

Laroche, C., & Michaud, P. (2007). New developments and prospective applicataions for β (1,3) glucans. *Recent Patents on Biotechnology*, *1*, 59–73.

Latgé, J.-P. (2007). The cell wall: A carbohydrate armour for the fungal cell. *Molecular Microbiology*, *66*, 279–290.

Latgé, J.-P. (2009). Galactofuranose containing molecules in *Aspergillus fumigatus*. *Medical Mycology*, *47*, S104–109. http://dx.doi.org/10.1080/1369378082258832.

Latgé, J.-P., Mouya, I., Tekaia, F., Beauvais, A., Debeaupuis, J. P., & Nierman, W. (2005). Specific molecular features in the organization and biosynthesis of the cell wall of *Aspergillus fumigatus*. *Medical Mycology*, *43*, S15–S22.

Lee, J. I., Choi, J. H., Park, B. C., Park, Y. H., Lee, M. Y., Park, H.-M., et al. (2004). Differential expression of the chitin synthase genes of *Aspergillus nidulans*, *chsA*, *chsB*, and *chsC*, in response to developmental status and environmental factors. *Fungal Genetics and Biology*, *41*, 635–646.

Lehle, L., Eiden, A., Lehnert, K., Haselbeck, A., & Kopetzki, E. (1995). Glycoprotein biosynthesis in *Saccharomyces cerevisiae*: *ngd29*, an N-glycosylation mutant allelic to *och1* having a defect in the initiation of outer chain formation. *FEBS Letters*, *370*, 41–45.

Leitão, E. A., Bittencourt, V. C. B., Haido, R. M. T., Valente, A. P., Peter-Katalinic, J., Letzel, M., et al. (2003). β-Galactofuranose-containing O-linked oligosaccharides present in the cell wall peptidogalactomannan of *Aspergillus fumigatus* contain immunodominant epitopes. *Glycobiology, 13*, 681–692.

Lenardon, M. D., Milne, S. A., Mora-Montes, H. M., Kaffarnik, F. A. R., Peck, S. C., Brown, A. J. P., et al. (2010). Phosphorylation regulates polarization of chitin synthesis in *Candida albicans*. *Journal of Cell Science, 123*, 2199–2206.

Lerch, K. (1983). Neurospora tyrosinase: Structural, spectroscopic and catalytic properties. *Molecular and Cellular Biochemistry, 52*, 125–138.

Lesage, G., & Bussey, H. (2006). Cell wall assembly in *Saccharomyces cerevisiae*. *Microbiology and Molecular Biology Reviews, 70*, 317–343.

Levin, D. E. (2011). Regulation of cell wall biogenesis in *Saccharomyces cerevisiae*: The cell wall integrity signaling pathway. *Genetics, 189*, 1145–1175.

Li, H., Zhou, H., Luo, Y., Ouyang, H., Hu, H., & Jin, C. (2007). Glycosylphosphatidylinositol (GPI) anchor is required in *Aspergillus fumigatus* for morphogenesis and virulence. *Molecular Microbiology, 64*, 1014–1027.

Liu, Y., & Filler, S. G. (2011). *Candida albicans* Als3, a multifunctional adhesion and invasion. *Eukaryotic Cell, 10*, 168–173.

Liu, J., Wang, H., McCollum, D., & Balasubramanian, M. K. (1999). Drc1p/Cps1p, a 1,3-β-glucan synthase subunit, is essential for division septum assembley in *Schizosaccharomyces pombe*. *Genetics, 153*, 1193–1203.

Lu, C. F., Montijn, R. C., Brown, J. L., Klis, F. M., Kurjan, J., Bussey, H., et al. (1995). Glycosylphosphatidylinositol-dependent cross-linking of alpha-agglutinin and beta 1,6-glucan in the *Saccharomyces cerevisiae* cell wall. *The Journal of Cell Biology, 128*, 333–340.

Maddi, A., Bowman, S. M., & Free, S. J. (2009). Trifluoromethanesulfonic acid-based proteomic analysis of cell wall and secreted proteins of the ascomycetous fungi *Neurospora crassa* and *Candida albicans*. *Fungal Genetics and Biology, 46*, 768–781.

Maddi, A., Dettmann, A., Fu, C., Seiler, S., & Free, S. J. (2012). WSC-1 and HAM-7 are MAK 1 MAP kinase pathway sensors required for cell wall integrity and cell fusion in *Neurospora crassa*. *PLoS One, 7*, e42374.

Maddi, A., & Free, S. J. (2010). α-1,6-Mannosylation of N-linked oligosaccharide present on cell wall proteins is required for their incorporation into the cell wall in the filamentous fungus *Neurospora crassa*. *Eukaryotic Cell, 9*, 1766–1775.

Maddi, A., Fu, C., & Free, S. J. (2012). The *Neurospora crassa dfg5* and *dcw1* genes encode α-1,6-mannanases that function in the incorporation of glycoproteins into the cell wall. *PLoS One, 7*, e38872.

Magnelli, P. E., Cipollo, J. F., & Robbins, P. W. (2005). A glucanase-driven fractionation allows redefinition of *Schizosaccharomyces pombe* cell wall composition and structure: Assignment of diglucan. *Analytical Biochemistry, 336*, 202–212.

Martinez-Lopez, R., Monteoliva, L., Diez-Orejas, R., Nombela, C., & Gil, C. (2004). The GPI anchored protein CaEcm33p is required for cell wall integrity, morphogenesis and virulence in *Candida albicans*. *Microbiology, 150*, 3341–3354.

Martinez-Lopez, R., Park, H., Myers, C. L., Gil, C., & Filler, S. G. (2006). *Candida albicans* Ecm33p is important for normal cell wall architecture and interactions with host cells. *Eukaryotic Cell, 5*, 140–147.

Martin-Garcia, R., Durán, A., & Valdivieso, M.-H. (2003). In *Schizosaccharomyces pombe* chs2p has no chitin synthase activity but is related to septum formation. *FEBS Letters, 549*, 176–180.

Matsuo, Y., Tanaka, K., Nakagawa, T., Matsuda, H., & Kawamukai, M. (2004). Genetic analysis of chs1^{+} and chs2^{+} encoding chitin synthases from *Schizosaccharomyces pombe*. *Bioscience, Biotechnology, and Biochemistry, 68*, 1489–1499.

Maubon, D., Park, S., Tanguy, M., Huerre, M., Schmitt, C., Prevost, M. C., et al. (2006). AGS3, an α-(1,3)glucan synthase gene family member of *Aspergillus fumigatus*, modulates mycelium growth in the lung of experimentally infected mice. *Fungal Genetics and Biology*, *43*, 366–375.

Mazur, P., & Baginsky, W. (1996). In vitro activity of 1,3-β-D-glucan synthase requires the GTP-binding protein Rho1. *The Journal of Biological Chemistry*, *271*, 14604–14609.

Mellado, C., Dubreucq, G., Mol, P., Sarfati, J., Paris, S., Diaquin, M., et al. (2003). Cell wall biogenesis in a double chitin synthase mutant (*chsG⁻/chsE⁻*) of *Aspergillus fumigatus*. *Fungal Genetics and Biology*, *38*, 98–109.

Merkel, G. J., & Scofield, B. A. (1997). The in vitro interaction of *Cryptococcus neoformans* with human lung epithelial cells. *FEMS Immunology and Medical Microbiology*, *19*, 203–213.

Miermont, A., Uhlendorf, J., McClean, M., & Hersen, P. (2011). The dynamic system properties of the HOG signaling cascade. *Journal of Signal Transduction*, *2011*, http://dx.doi.org/10.1155/2011/930940.

Mio, T., Yabe, T., Sudoh, M., Satoh, Y., Nakamina, T., Arisawa, M., et al. (1996). Role of three chitin synthase genes in the growth of *Candida albicans*. *Journal of Bacteriology*, *178*, 2416–2419.

Monod, M., Hube, B., Hess, D., & Sanglard, D. (1998). Differential regulation of SAP8 and SAP9, which encode two members of the secreted aspartyl proteinase family in *Candida albicans*. *Microbiology*, *144*, 2731–2737.

Montijn, R. C., Vink, E., Müller, W. H., Verkleij, A. J., Van Den Ende, H., Henrissat, B., et al. (1999). Localization of synthesis of β1,6-glucan in *Saccharomyces cerevisiae*. *Journal of Bacteriology*, *181*, 7414–7420.

Morris-Jones, R., Gomez, B. L., Diez, S., Uran, M., Morris-Jones, S. D., Casadevall, A., et al. (2005). Synthesis of melanin pigment by *Candida albicans* in vitro and during infection. *Infection and Immunity*, *73*, 6147–6150.

Mouyna, I., Fontaine, T., Vai, M., Monod, M., Fonzi, W. A., Diaquin, M., et al. (2000). Glycosylphosphatidylinositol-anchored glucanosyltransferases play an active role in the biosynthesis of the fungal cell wall. *The Journal of Biological Chemistry*, *275*, 14882–14889.

Mouyna, I., Morelle, W., Vai, M., Monod, M., Léchenne, B., Fontaine, T., et al. (2005). Deletion of GEL2 encoding for a β(1,3)glucanosyltransferase affects morphogenesis and virulence in *Aspergillus fumigatus*. *Molecular Microbiology*, *56*, 1675–1688.

Mrsa, V., & Tanner, W. (1999). Role of NaOH-extractable cell wall proteins Ccw5p, Ccw6p, Ccw7p and Ccw8p (Members of the Pir protein family) in stability of the *Saccharomyces cerevisiae* cell wall. *Yeast*, *15*, 813–820.

Munro, C. A., Whitton, R. K., Hughes, H. B., Rella, M., Selvaggini, S., & Gow, N. A. R. (2003). CHS8—A fourth chitin synthase gene of *Candida albicans* contributes to the in vitro chitin synthase activity, but is dispensable for growth. *Fungal Genetics and Biology*, *40*, 146–158.

Munro, C. A., Winter, K., Buchan, A., Henry, K., Becker, J. M., Brown, A. J. P., et al. (2001). Chs1 of *Candida albicans* is an essential chitin synthase required for synthesis of the septum and for cell integrity. *Molecular Microbiology*, *39*, 1414–1426.

Nakajima, T., Yoshida, M., Nakamura, M., Hura, N., & Matsuda, K. (1984). Structure of the cell wall proteogalactomannan from *Neurospora crassa* II: Structural analysis of the polysaccharide part. *Journal of Biochemistry*, *96*, 1013–1020.

Nakanishi-Shindo, Y., Nakayama, K., Tanaka, A., Toda, Y., & Jigami, Y. (1993). Structure of the N-linked oligosaccharides that show the complete loss of α-1,6-polymannose outer chain from *och1*, *och1 mnn1*, and *mnn1 alg3* mutants of *Saccharomyces cerevisiae*. *The Journal of Biological Chemistry*, *268*, 26338–26345.

Nobbs, A. H., Vickerman, M. M., & Jenkinson, H. F. (2010). Heterologous expression of *Candida albicans* cell wall-associated adhesins in *Saccharomyces cerevisiae* reveals differential specificities in adherence and biofilm formation and in binding oral *Streptococcus gordonii*. *Eukaryotic Cell*, *9*, 1622–1634.

Nobile, C. J., Nett, J. E., Andes, D. R., & Mitchell, A. P. (2006). Function of *Candida albicans* adhesin Hwp1p in biofilm formation. *Eukaryotic Cell*, *5*, 1604–1610.

Norice, C. T., Smith, F. J., Solis, N., Filler, S. G., & Mitchell, A. P. (2007). Requirement for *Candida albicans* Sun41 in biofilm formation and virulence. *Eukaryotic Cell*, *6*, 2046–2055.

Olson, G. M., Fox, D. S., Wang, P., Alspaugh, J. A., & Buchanan, K. L. (2007). Role of protein O-mannosyltransferase Pmt4 in the morphogenesis and virulence of *Cryptococcus neoformans*. *Eukaryotic Cell*, *6*, 222–234.

Pardini, G., de Groot, P. W. J., Coste, A. T., Karababa, M., Klis, F. M., de Koster, C. G., et al. (2006). The CRH family coding for cell wall glycosylphosphatidylinositol proteins with a predicted transglycosidase domain affects cell wall organization and virulence of *Candida albicans*. *The Journal of Biological Chemistry*, *281*, 40399–40411.

Pardo, M., Monteoliva, L., Vázquez, P., Martínez, M. R., Molero, G., Nombela, C., et al. (2004). PST1 and ECM33 encode two yeast cell surface GPI proteins important for cell wall integrity. *Microbiology*, *150*, 4157–4170.

Pardo, M., Ward, M., Bains, S., Molina, M., Blackstock, W., Gil, C., et al. (2000). A proteomic approach for the study of *Saccharomyces cerevisiae* cell wall biogenesis. *Electrophoresis*, *21*, 3396–3410.

Park, J. N., Lee, D. J., Kwon, O., Oh, D. B., Bahn, Y. S., & Kang, H. A. (2012). Unraveling unique structure and biosynthesis pathway of N-linked glycans in human fungal pathogen *Cryptococcus neoformans* by glycomics analysis. *The Journal of Biological Chemistry*, *287*, 19501–19515.

Pitarch, A., Sánchez, M., Nombela, C., & Gil, C. (2002). Sequential fractionation and two-dimensional gel analysis unravels the complexity of the dimorphic fungus *Candida albicans* cell wall proteome. *Molecular and Cellular Proteomics*, *1*, 967–982.

Plaine, A., Walker, L., Da Costa, G., Mora-Montes, H. M., McKinnon, A., Gow, N. A. R., et al. (2008). Functional analysis of *Candida albicans* GPI-anchored proteins: Roles in cell wall integrity and caspofungin sensitivity. *Fungal Genetics and Biology*, *45*, 1404–1414.

Rademaker, G. J., Pergantis, S. A., Blok-Tip, L., Langridge, J. I., Kleen, A., & Thomas-Oates, J. E. (1998). Mass spectrometric determination of the sites of O-glycan attachment with low picomolar sensitivity. *Analytical Biochemistry*, *257*, 149–160.

Ragni, E., Coluccio, A., Rolli, E., Rodriguez-Peña, J. M., Colasante, G., Arroyo, J., et al. (2007). GAS2 and GAS4, a pair of developmentally regulated genes required for spore wall assembly in *Saccharomyces cerevisiae*. *Eukaryotic Cell*, *6*, 302–316.

Ragni, E., Fontaine, T., Gissi, C., Latgé, J. P., & Popolo, L. (2007). The Gas family of proteins of *Saccharomyces cerevisiae*: Characterization and evolutionary analysis. *Yeast*, *24*, 297–308.

Ram, A. F., Kapteyn, J. C., Montijn, R. C., Caro, L. H., Douwes, J. E., Baginsky, W., et al. (1998). Loss of the plasma membrane-bound protein Gas1p in *Saccharomyces cerevisiae* results in the release of beta1,3-glucan into the medium and induces a compensation mechanism to ensure cell wall integrity. *Journal of Bacteriology*, *180*, 1418–1424.

Reese, A. J., Yoneda, A., Breger, J. A., Beauvais, A., Liu, H., Griffith, C. L., et al. (2007). Loss of cell wall alpha(1,3)glucan affects *Cryptococcus neoformans* from ultrastructure to virulence. *Molecular Microbiology*, *63*, 1385–1398.

Richard, M. (2002). Complete glycosylphosphatidylinositol anchors are required in *Candida albicans* for full morphogenesis, virulence and resistance to macrophages. *Molecular Microbiology*, *44*, 841–853.

Roetzer, A., Gabaldón, T., & Schüller, C. (2011). From *Saccharomyces cerevisiae* to *Candida glabrata* in a few easy steps: Important adaptations for an opportunistic pathogen. *FEMS Microbiology Letters*, *314*, 1–9.

Rolli, E., Ragni, E., De Medina-Redondo, M., Arroyo, J., De Aldana, C. R., & Popolo, L. (2011). Expression, stability, and replacement of glucan-remodeling enzymes during developmental transitions in *Saccharomyces cerevisiae*. *Molecular Biology of the Cell*, *22*, 1585–1598.

Romano, J., Nimrod, G., Ben-Tal, N., Shadkchan, Y., Baruch, K., Sharon, H., et al. (2006). Disruption of the *Aspergillus fumigatus* ECM33 homologue results in rapid conidial germination, antifungal resistance and hypervirulence. *Microbiology*, *152*, 1919–1928.

Roncero, C. (2002). The genetic complexity of chitin synthesis in fungi. *Current Genetics*, *41*, 367–378.

Ruiz-Herera, J., & San-Blas, G. (2003). Chitin synthesis as a target for antifungal drugs. *Current Drug Targets. Infectious Disorders*, *3*, 77–91.

Ruiz-Herrera, J., Elorza, M. V., Valentin, E., & Sentandreu, R. (2006). Molecular organization of the cell wall of *Canidida albicans* and its relation to pathogenicity. *FEMS Yeast Research*, *6*, 14–29.

Ruiz-Herrera, J., González-Prieto, J. M., & Ruiz-Medrano, R. (2002). Evolution and phylogenetic relationships of chitin synthases from yeasts and fungi. *FEMS Yeast Research*, *1*, 247–256.

Sánchez-León, E., Verdín, J., Freitag, M., Roberson, R. W., Bartnicki-Garcia, S., & Riquelmi, M. (2011). Traffic of chitin synthase 1 (CHS-1) to the Spitzenkörper and developing septa in hyphae of *Neurospora crassa*: Actin dependence and evidence of distinct microvesicle populations. *Eukaryotic Cell*, *10*, 683–695.

Sarthy, A. V., McGonigal, T., Coen, M., Frost, D. J., Meulbroek, J. A., & Goldman, R. C. (1997). Phenotype in *Candida albicans* of a disruption of the BGL2 gene encoding a 1,3-teta-glucosyltransferase. *Microbiology*, *143*, 367–376.

Sestak, S., Hagen, I., Tanner, W., & Strahl, S. (2004). Scw10p, a cell wall glucanase/transglucosidase important for cell wall stability in *Saccharomyces cerevisiae*. *Microbiology*, *150*, 3197–3208.

Sharifmoghadam, M. R., & Valdivieso, M. H. (2008). The *Schizosaccharomyces pombe* Map4 adhesin is a glycoprotein that can be extracted from the cell wall with alkali but not with beta-glucanases and requires the C-terminal DIPSY domain for function. *Molecular Microbiology*, *69*, 1476–1490.

Sheppard, D. C. (2011). Molecular mechanism of *Aspergillus fumigatus* adherence to host constituents. *Current Opinion in Microbiology*, *14*, 375–379.

Sosinska, G. J., de Koning, L. J., de Groot, P. W. J., Manders, E. M. M., Dekker, H. L., Hellingwerf, K. J., et al. (2011). Mass spectrometric quantification of the adaptations in the wall proteome of *Candida albicans* in response to ambient pH. *Microbiology*, *157*, 136–146.

Spreghini, E., Davis, D. A., Subaran, R., Kim, M., & Mitchell, A. P. (2003). Roles of *Candida albicans* Dfg5p and Dcw1p cell surface proteins in growth and hypha formation. *Eukaryotic Cell*, *2*, 746–755.

Staab, J. F., Bradway, S. D., Fidel, P. L., & Sundstrom, P. (1999). Adhesive and mammalian transglutaminase substrate properties of *Candida albicans* Hwp1. *Science*, *283*, 1535–1538.

Stanley, P., Schachter, H., & Taniguchi, N. (2009). Chapter 8. N-glucans. In A. Varki, R. D. Cummings, J. D. Esko, H. H. Freeze, P. Stanley, C. R. Bertozzi, G. W. Hart & M. E. Etzler (Eds.), *Essentials of glycobiology* (pp. 101–114). Cold Spring Harbor, NY: Cold Spring Harbor Press.

Sugareva, V., Härtl, A., Brock, M., Hübner, K., Rohde, M., Heinekamp, T., et al. (2006). Characterization of the laccase-encoding gene abr2 of the dihydroxynapthalene-like melanin gene cluster of *Aspergillus fumigatus*. *Archives of Microbiology*, *186*, 345–355. http://dx.doi.org/10.1007/s00203-006-0144-2.

Sundstrom, P. (2002). Adhesion in *Candida* spp. *Cellular Microbiology*, *4*, 461–469.

Taib, M., Pinney, J. W., Westhead, D. R., McDowall, K. J., & Adams, D. J. (2005). Differential expression and extent of fungal/plant and fungal/bacterial chitinases of *Aspergillus fumigatus*. *Archives of Microbiology*, *184*, 78–81.

Takada, H., Nishida, A., Domae, M., Kita, A., Yamano, Y., Uchida, A., et al. (2010). The cell surface protein gene *ecm33+* is a target of the two transcription factors Atf1 and Mbx1 and negatively regulates Pmk1 MAPK cell integrity signaling in fission yeast. *Molecular Biology of the Cell*, *21*, 674–685.

Tentler, S., Palas, J., Enderlin, C., Campbell, J., Taft, C., Miller, T. K., et al. (1997). Inhibition of *Neurospora crassa* growth by a glucan synthase-1 antisense construct. *Current Microbiology*, *34*, 303–308.

Teparic, R., Stuparevic, I., & Mrsa, V. (2004). Increased mortality of *Saccharomyces cerevisiae* cell wall protein mutants. *Microbiology*, *150*, 3145–3150.

Thompson, J. R., Douglas, C. M., Li, W., Jue, C. K., Pramanik, B., Yuan, X., et al. (1999). A glucan synthase FKS1 homolog in *Cryptococcus neoformans* is single copy and encodes an essential function. *Journal of Bacteriology*, *181*, 444–453.

Tsai, H.-F., Wheeler, M. H., Chang, Y. C., & Kwon-Chung, K. J. (1999). A developmentally regulated gene cluster involved in conidial pigment biosynthesis in *Aspergillus fumigatus*. *Journal of Bacteriology*, *181*, 6469–6477.

Upadhyay, S. K., Mahajan, L., Ramjee, S., Basir, S. F., & Madan, T. (2009). Identification and characterization of a laminin-binding protein of *Aspergillus fumigatus*: Extracellular thaumatin-domain protein (AfCalAp). *Journal of Medical Microbiology*, *58*, 714–722.

Utsugi, T., Minemura, M., Hirata, A., Abe, M., Watanabe, D., & Ohya, Y. (2002). Movement of yeast 1,3-β-glucan synthase is essential for uniform cell wall synthesis. *Genes to Cells*, *7*, 1–9.

Vadaie, N., Dionne, H., Akajagbor, D. S., Nickerson, S. R., Krysan, D. J., & Cullen, P. J. (2008). Cleavage of the signaling mucin Msb2 by the aspartyl protease Yps1 is required for MAPK activation in yeast. *The Journal of Cell Biology*, *181*, 1073–1081.

Verdín, J., Bartnicki-Garcia, S., & Riquelme, M. (2009). Functional stratification of the Spitzenkörper of *Neurospora crassa*. *Molecular Microbiology*, *74*, 1044–1053.

Verstrepen, K. J., & Klis, F. M. (2006). Flocculation, adhesion and biofilm formation in yeasts. *Molecular Microbiology*, *60*, 5–15.

Victoria, G. S., Kumar, P., & Komath, S. S. (2010). The *Candida albicans* homologue of PIG-P, CaGpi19p: Gene dosage and role in growth and filamentation. *Microbiology*, *156*, 3041–3051.

Wagener, J., Echtenacher, B., Rohde, M., Kotz, A., Krappmann, S., Heesemann, J., et al. (2008). The putative α-1,2-mannosyltransferase AfMnt1 of the opportunistic fungal pathogen *Aspergillus fumigatus* is required for cell wall stability and full virulence. *Eukaryotic Cell*, *7*, 1661–1673.

Walker, C. A., Gómez, B. L., Mora-Montes, H. M., Mackenzie, K. S., Munro, C. A., Brown, A. J. P., et al. (2010). Melanin externalization in *Candida albicans* depends on cell wall chitin structures. *Eukaryotic Cell*, *9*, 1329–1342.

Wang, Y., Aisen, P., & Casadevall, A. (1995). *Cryptococcus neoformans* melanin and virulence: Mechanism of action. *Infection and Immunity*, *63*, 3131–3136.

Wang, L., Zhai, B., & Lin, X. (2012). The link between morphotype transition and virulence in *Cryptococcus neoformans*. *PLoS Pathogens*, *8*, e1002765.

Williamson, P. R. (1997). Laccase and melanin in the pathogenesis of *Cryptococcus neoformans*. *Frontiers in Bioscience*, *2*, 99–107.

Xia, G., Jin, C., Zhou, J., Yang, S., Zhang, S., & Jin, C. (2001). A novel chitinase having a unique mode of action from *Aspergillus fumigatus* YJ-407. *European Journal of Biochemistry*, *268*, 4079–4085.

Yin, Q. Y., de Groot, P. W. J., de Koster, C. G., & Klis, F. M. (2007). Mass spectrometry-based proteomics of fungal wall glycoproteins. *Trends in Microbiology*, *16*, 20–25.

Yin, Q. Y., de Groot, P. W. J., Dekker, H. L., de Jong, L., Klis, F. M., & de Koster, C. G. (2005). Comprehensive proteomic analysis of *Saccharomyces cerevisiae* cell walls. *The Journal of Biological Chemistry*, *280*, 20894–20901.

Youngchim, S., Morris-Jones, R., Haly, R. J., & Hamilton, A. J. (2004). Production of melanin by *Aspergillus fumigatus*. *Journal of Medical Microbiology*, *53*, 175–181.

Yun, D.-J., Zhao, Y., Pardo, J. M., Narasimhan, M. L., Damsz, D., Lee, H., et al. (1997). Stress proteins on the yeast cell surface determine resistance to osmotin, a plant antifungal protein. *Proceedings of the National Academy of Sciences of the United States of America*, *94*, 7082–7087.

Zhao, X., Daniels, K. J., Oh, S.-H., Green, C. B., Yeater, K. M., Soll, D. R., et al. (2006). *Candida albicans* Als3p is required for wild-type biofilm formation on silicone elastomer surfaces. *Microbiology*, *152*, 2287–2299.

Zhao, X., Oh, S.-H., Yeater, K. M., & Hoyer, L. L. (2005). Analysis of the *Candida albicans* Als2p and Als4p adhesins suggests the potential for compensatory function within the Als family. *Microbiology*, *151*, 1619–1630.

Zhou, H., Hu, H., Zhang, L., Li, R., Ouyang, H., Ming, J., et al. (2007). O-Mannosyltransferase 1 in *Aspergillus fumigatus* (AfPmt1p) is crucial for cell wall integrity and conidium morphology, especially at an elevated temperature. *Eukaryotic Cell*, *6*, 2260–2268.

CHAPTER THREE

Polycomb Group Response Elements in *Drosophila* and Vertebrates

Judith A. Kassis[1], J. Lesley Brown

Eunice Kennedy Shriver National Institute of Child Health and Human Development, National Institutes of Health, Bethesda, Maryland, USA
[1]Corresponding author: e-mail address: jkassis@mail.nih.gov

Contents

Abstract

Polycomb group genes (PcG) encode a group of about 16 proteins that were first identified in *Drosophila* as repressors of homeotic genes. PcG proteins are present in all metazoans and are best characterized as transcriptional repressors. In *Drosophila*, these proteins are known as epigenetic regulators because they remember, but do not establish, the patterned expression state of homeotic genes throughout development. PcG proteins, in general, are not DNA binding proteins, but act in protein complexes to repress transcription at specific target genes. How are PcG proteins recruited to the DNA? In *Drosophila*, there are specific regulatory DNA elements called Polycomb group response elements (PREs) that bring PcG protein complexes to the DNA. *Drosophila* PREs are made up of binding sites for a complex array of DNA binding proteins. Functional PRE assays in transgenes have shown that PREs act in the context of other regulatory DNA and PRE activity is highly dependent on genomic context. *Drosophila* PREs tend to regulate genes with a complex array of regulatory DNA in a cell or tissue-specific fashion and it is the interplay between regulatory DNA that dictates PRE function. In mammals, PcG proteins are more diverse and there are multiple ways to recruit PcG complexes, including RNA-mediated recruitment. In this review, we discuss evidence for PREs in vertebrates and explore similarities and differences between *Drosophila* and vertebrate PREs.

1. INTRODUCTION

Polycomb group proteins (PcG) are important regulators of developmental genes in all metazoans (two reviews: Beisel & Paro, 2011; Simon & Kingston, 2009). First discovered in genetic studies as regulators of homeotic genes in *Drosophila*, genome-wide chromatin-immunoprecipitation (ChIP) studies using antibodies against various PcG proteins show that there are likely hundreds of PcG targets in *Drosophila* (Négre et al., 2006; Schwartz et al., 2006, 2010; Tolhuis et al., 2006). In vertebrate embryonic stem (ES) cells, PcG proteins are important regulators of developmental genes and are important in both maintenance of pluripotency and in differentiation. Because they regulate developmental and cell cycle genes, altered expression of *PcG* genes has been associated with many cancers (reviews: Geini & Hendzel, 2009; Sparmann & van Lohuizen, 2006).

There are at least 16 *PcG* genes in *Drosophila*, identified by mutations that derepress homeotic gene expression. Many *PcG* genes encode proteins that act in protein complexes to regulate transcription via alterations to chromatin structure. The two best studied PcG protein complexes are PRC1 and PRC2 (Polycomb repressive complexes 1 and 2), which are conserved from flies to mammals (for reviews, see Kerppola, 2009; Müller & Verrijzer, 2009; Schuettengruber & Cavalli, 2009; Simon & Kingston, 2009). PRC2 contains the PcG proteins Enhancer of zeste (E(z)), Extra sex combs (Esc), Suppressor of zeste 12 (Su(z)12), as well as the protein p55 (Cao et al., 2002; Czermin et al., 2002; Müller et al., 2002), and in some tissues Polycomb-like (Pcl), which modifies PRC2 function (Nekrasov et al., 2007; Savla, Benes, Zhang, & Jones, 2008). E(z), the catalytic component of PRC2, trimethylates histone H3 lysine 27 creating the H3K27me3 mark characteristic of Polycomb-regulated genes. The PRC1 core complex is comprised of Polycomb (Pc), Polyhomeotic (Ph), Posterior sex combs (Psc), and Sex combs extra/dRing1 (Sce/dRing1; Fritsch, Beuchle, & Müller, 2003) and acts to inhibit chromatin remodeling and compact chromatin (Saurin, Shao, Erdjument-Bromage, Tempst, & Kingston, 2001; Shao et al., 1999). Psc and Sce/dRing1 are also in another complex called dRAF that contains a histone demethylase, dKDM2 (Lagarou et al., 2008). Another PcG complex, PR-DUB (Polycomb repressive deubiquitinase) consists of two other PcG proteins, Calypso and Additional sex combs (Asx), which deubiquitinates H2A118 (Scheuermann et al., 2010; Schuettengruber & Cavalli, 2010). Additional PcG proteins required for homeotic gene silencing not yet assigned to a protein complex include Sex combs on middle leg (Scm), which loosely associates with PRC1, and Super sex combs (Sxc/Ogt) that encodes O-GlcNAc transferase and modifies Ph (Gambetta, Oktaba, & Müller, 2009). Finally, *Drosophila* has the protein complex Pho-repressive complex (PhoRC), consisting of the DNA binding PcG protein Pleiohomeotic (Pho) and the methyl-lysine-binding protein, dSfmbt (Klymenko et al., 2006). Both Pho and dSfmbt have homologues in vertebrates, but no vertebrate PhoRC complex has been described.

Genome-wide studies have shown that PcG proteins colocalize to many or most PcG target genes in *Drosophila*. Despite this, mutations in various *PcG* genes give very different phenotypes (Breen & Duncan, 1986). This suggests that different target genes have different requirements for specific PcG proteins. For example, at the PcG target gene *engrailed* (*en*), a gene required for segmentation, mutations in *ph* cause a massive *en* derepression in embryos, while mutations in *Pc* cause very little *en* misexpression (Moazed & O'Farrell, 1992). Since Ph and Pc are both components of PRC1, and both are bound to *en* PREs in *Drosophila* embryos (Négre et al., 2006), why *en* is so sensitive to the loss of Ph but not Pc is unclear. In a recent study, Gutiérrez et al. (2012) began to classify different PcG targets based on their response to *Sce* mutations. They studied targets that bind the PcG proteins Pho, Sce, Pc, Ph, and Psc, and found that expression of only a subset of the targets was altered in *Sce* mutants. Clearly, there is still a lot we need to learn about the role of different PcG proteins in transcriptional repression of different target genes.

In this review, we focus on the fundamental question of how PcG proteins are recruited to their target genes. This topic has been comprehensively covered in a number of recent reviews (Beisel & Paro, 2011; Müller & Kassis, 2006; Ringrose & Paro, 2007; Schuettengruber & Cavalli, 2009); here, we bring our own perspective and discuss unresolved issues. In *Drosophila*, we concentrate on target genes that are bound by both PRC1 and PRC2 and describe what is known about Polycomb group response element (PREs),

DNA fragments that can recruit PRC1 and PRC2 to target genes via a complex array of DNA binding proteins. In vertebrates, recruitment of PcG proteins is more complicated, given that there are many more PcG proteins and varieties of PcG protein complexes (Beisel & Paro, 2011; Kerppola, 2009). While it has been demonstrated that RNA is involved in recruitment of PcG protein complexes at some vertebrate PcG targets, a role for RNA in PcG protein binding in *Drosophila* has not been demonstrated. Thus, for the vertebrate system, we limit our discussion to presumptive PREs.

2. PREs IN *DROSOPHILA*

2.1. PRE maintenance assays

PREs were first discovered in transgenes of regulatory DNA from the Bithorax complex (BX-C). In 1991, Müller and Bienz, while studying regulatory DNA from the *Ubx* gene, identified a DNA fragment that they called BXD, which, in combination with another regulatory fragment that initiated correct *Ubx* expression, could maintain β-galactosidase expression throughout embryonic development in a pattern similar to *Ubx*. Notably, in a *Polycomb* mutant, β-galactosidase expression was derepressed late in embryonic development, suggesting that *Polycomb* directly repressed the expression of the transgene (Müller & Bienz, 1991). In 1993, Simon et al. identified three discrete fragments of regulatory DNA from the BX-C that rendered transgenes responsive to PcG repression and coined the term Polycomb group response elements (PREs). Several important principles came out of this and other work on PREs from the BX-C complex. First, PREs act with other regulatory DNA to remember boundaries set up directly by gap and pair-rule regulators. Second, PREs respond to mutations in many different *PcG* genes, suggesting PcG proteins act together to repress gene expression. Third, discrete BX-C PREs have similar activities and can be interchangeable within transgenes (Simon, Chiang, Bender, Shimell, & O'Connor, 1993).

The principles of PREs learned from the early experiments on BX-C PREs apply to most if not all of the PREs studied to date. We study PREs from the *Drosophila engrailed* (*en*) and *invected* (*inv*) genes. *en* and *inv* encode coregulated, highly related homeodomain proteins which function together during development to control embryonic segmentation and formation of the posterior compartment in imaginal disks (Gustavson, Goldsborough, Ali, & Kornberg, 1996). *en* and *inv* exist in a 115-kb chromatin domain, marked by the distinctive H3K27me3 modification put on by the PRC2 PcG protein complex (Kharchenko et al., 2011). There are four major PREs present in the *en/inv* domain (Cunningham, Brown, & Kassis, 2010; DeVido, Kwon, Brown, & Kassis, 2008) and their activity can be demonstrated in the functional assay shown in Fig. 3.1. A transgene containing 8 kb of upstream *en* regulatory DNA fused to the reporter gene *lacZ* expresses β-galactosidase in *en*-like stripes throughout embryonic development. However, deletion of a 2-kb fragment, that includes two PREs, while still giving *en*-like stripes early in development, gives expression throughout the embryo in late development. Thus, the 2-kb fragment, containing two PREs, is required for maintenance of *en* stripes. In this construct, the activity of both PREs is required for full maintenance (DeVido et al., 2008). Interestingly, the *en* PREs are also able to act as PREs in an embryonic *Ubx*-reporter construct (Americo et al., 2002) and PRE$_D$, a PRE from the *Ubx* gene, can act with *en* regulatory DNA to maintain *en*-expression in stripes (Cunningham et al., 2010). These data emphasize the functional similarity between *en* and *Ubx* PREs.

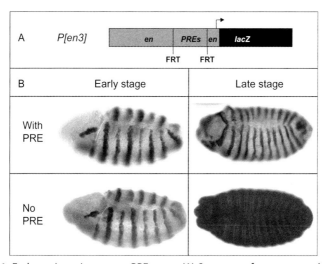

Figure 3.1 Embryonic maintenance PRE assay. (A) Structure of a vector used to test for PRE activity of the *en* PREs. Eight kilobases of *en* sequence is fused to the reporter gene, *lacZ*. This leads to the expression of β-galactosidase in *en*-like stripes throughout embryonic development. This 8-kb of *en* includes a 2-kb region that encompasses the *en* PREs. Here, the 2-kb fragment containing the PREs is flanked by FRT sites allowing for removal of the PRE with FLP recombinase (Cunningham et al., 2010; DeVido et al., 2008). (B) Early and late stage embryos stained for β-galactosidase with a line that carries the 2-kb *en* PRE and a line that has the 2 kb PRE excised. With the PRE fragment, *lacZ* expression is maintained in *en*-like stripes throughout embryonic development. When the *en* PRE is excised, *lacZ* is expressed throughout the embryo. Early stage embryos: lateral view, anterior left, dorsal up, stage 11. Late stage embryos: dorsal view, anterior left, stage 16.

In addition to embryonic reporter constructs, there are at least two imaginal disk *Ubx-lacZ*-reporter constructs for testing PRE activity (Fritsch, Brown, Kassis, & Müller, 1999; Poux, Kostic, & Pirrotta, 1996); both constructs contain combinations of *Ubx*-regulatory DNA, including an imaginal disk enhancer. Without a PRE, β-galactosidase is ubiquitously expressed in the wing disk from these constructs, where *Ubx* is not normally expressed; addition of a PRE represses wing disk expression of the *Ubx-lacZ*-reporter. An *Ubx-lacZ* imaginal disk reporter construct was used to refine the size of the *Ubx bxd* PRE, PRE_D to a 567-bp fragment and to demonstrate the importance of the PcG-DNA-binding protein Pho to PRE_D activity in larvae (Fritsch et al., 1999). Recently, an imaginal *Ubx-lacZ*-reporter was used to test the activity of presumptive PREs from the *Psc–Su(z)2* gene complex (Park, Schwartz, Kahn, Asker, & Pirrotta, 2012). These results again demonstrated that DNA from different genes can act as PREs with *Ubx*-regulatory DNA.

Only a few of the hundreds of PREs identified in genome-wide studies have been tested for activity in either embryonic or imaginal disk PRE reporter constructs and even fewer have been tested in both (see Table 3.1). It might be that some PREs are stage or tissue specific (see below). The development of a series of PRE vectors, with regulatory DNA from different genes, designed to assess activity at different developmental stages using the phi-C31 system (Groth, Fish, Nusse, & Calos, 2004) to insert transgenes at the same chromosomal location, would be key to assessing the functional equivalency between PREs.

2.2. Mini-*white* silencing

The discovery of silencing of the mini-*white* gene in transgenic *Drosophila* coincided with the discovery of PREs (Kassis, Vansickle, & Sensabaugh, 1991). In 1988, Pirrotta developed *pCaSpeR*, a vector for P-element transformation using the mini-*white* gene as a reporter for transgenic flies (Pirrotta, 1988). *white* encodes a member of the ABC family of transporter superfamily and is necessary for the transport of the precursors of eye color pigment into the eye. Flies with a null mutation in *white* have a white-eye color. To make transgenic flies with the *pCaSpeR* vector, embryos mutant for the *white* gene are injected, mated, and screened for transgenic progeny that have colored eyes. The amount of *white* gene product produced determines the eye color; that is, a little White product gives a light yellow eye, more White product gives an orange eye all the way to the wild-type red eye

Table 3.1 Summary of *Drosophila* genes with PREs that have been characterized by the PSS assay or a maintenance assay in embryos (E) or larvae (L)

PRE	Maintenance assay	PSS mini-*white*
en	Yes, E	Yes
inv	Yes, E	Yes
eve	Yes, E	Yes
Psc/Su(z)2	Yes, L	Yes
Scr	Yes, E	Yes
pb	Yes, E	Yes
vg	ND	Yes
ph	ND	Yes
esg	ND	Yes
hh	ND	Yes
cad	ND	Yes
aPKC	ND	Yes
prod	ND	Yes
BX-C		
Mcp	Yes, E	Yes
bxd(PRE$_D$)	Yes, E, L	Yes
iab-2	Yes, E*	Yes
iab-6	ND	Yes
iab7 (Fab7)	Yes E	Yes
iab-8	Yes E	Yes

*iab-2 fragment used required additional DNA to work in the embryonic maintenance assay.
ND is for not done. Some of these genes have more than one PRE that have been tested in these assays (see text for discussion of this). *Scr*, Sex combs reduced; *aPKC*, atypical protein kinase C; *prod*, proliferation disruptor; *hh*, hedgehog; *vg*, vestigal; *cad*, caudal; *esg*, escargot; *pb*, probosipedia; *BX-C*, Bithorax complex. Mcp, bxd (PRE$_D$), iab fragments are regulatory regions within the BX-C.
References: *en*: Kassis et al. (1991), Kassis (1994), Brown, Mucci, Whiteley, Dirksen, and Kassis (1998), Americo et al. (2002), and DeVido et al. (2008). *inv*: Cunningham et al. (2010). *eve*: Fujioka, Emi-Sarker, Yusibova, Goto, and Jaynes (1999) and Fujioka, Yusibova, Zhou, and Jaynes (2008). *Psc/Su(z)2*: Park et al. (2012). *Scr*: Gindhart and Kaufman (1995). *pb*: Kapoun and Kaufman (1995). *vg*: Okulski, Druck, Bhalerao, and Ringrose (2011). *ph*: Bloyer et al. (2003). *esg*: Kassis (1994). *hh*: Rank, Prestel, and Paro (2002), Maurange and Paro (2002) and Chanas and Machat (2005). *cad*, *aPKC*, and *prod:* Ringrose, Rehmsmeier, Dura, and Paro (2003). *BX-C*: Chan, Rastelli, and Pirrotta (1994), Gruzdeva, Kyrchanova, Parshikov, Kullyev, and Georgiev (2005), Vazquez, Muller, Pirrotta, and Sedat (2006), Pérez-Lluch, Cuartero, Azorin, and Espinàs (2008), Mishra et al. (2001), Shimell, Peterson, Burr, Simon, and O'Connor (2000), and Barges et al. (2000).

color. The mini-*white* gene lacks most of the enhancers that cause *white* to be expressed in the eye; eyes from *pCaSpeR*-containing transgenic flies have an eye color ranging from yellow to red. This is because the expression of mini-*white* is highly susceptible to regulation by flanking genomic DNA. The *pCaSpeR* vector soon became the vector of choice because of two properties: (1) it was easier to clone into than other P-based vectors and (2) transgenic flies with independent insertion sites could be recognized by their different eye colors.

In 1991, Kassis et al. reported that a fragment of regulatory DNA from the *Drosophila en* gene could mediate mini-*white* silencing in flies homozygous for the *en*-mini-*white* containing transgenes (*P[en-mw]*). Normally, when using a vector with mini-*white*, flies homozygous for the transgene have a darker eye color than flies heterozygous for it (*P[mw]* in Fig. 3.2).

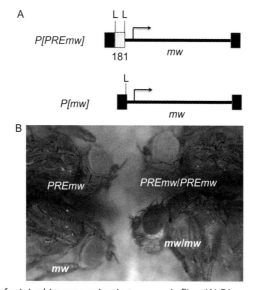

Figure 3.2 PSS of mini-*white* expression in transgenic flies. (A) Diagrammatic representation of the two reporter constructs used to test for PSS of an *en* PRE. *P[mw]* is the P-element vector control without any PRE insert. *P[PREmw]* is the same transformation vector carrying a 181-bp PRE from *en* cloned upstream of the mini-*white* promoter. The filled boxes represent the P-element ends. The open box represents the PRE. L is *loxP* sites that allow excision of the PRE. (B) Shows the eye color of transformants carrying the above constructs. *mw* and *PREmw* are heterozygous for the construct. *mw/mw* and *PREmw/PREmw* are homozygous for the constructs. PSS can be seen by comparing heterozygous versus homozygous eye colors for the two different constructs. *Figure adapted from Noyes, Stefaniuk, Cheng, Kennison, and Kassis (2011).* (See Color Insert.)

However, when a 2.4-kb fragment of *en* regulatory DNA was included in the mini-*white* vector, the eye color of homozygous flies was often lighter than that of heterozygous flies, sometimes even white, suggesting that the mini-*white* expression was completely silenced in the eye. This mini-*white* silencing was dependent on the two transgenes being inserted near each other in the genome, either in *cis* (a duplicated *P[en-mw]* element would also silence mini-*white*) or in *trans*, that is, on homologous chromosomes (Kassis et al., 1991). Somatic chromosomes are paired in *Drosophila*, and this mini-*white* silencing was dependent on the pairing of the two *P[en-mw]* transgenes. This phenomenon has been called pairing-dependent repression or pairing-sensitive silencing (PSS) (Kassis, 1994). The 2.4-kb *en* fragment contained within the *P[en-mw]* transgenes contains two PREs (Americo et al., 2002; DeVido et al., 2008; Kassis, 1994).

Mini-*white* silencing is a property of PREs (reviewed in Kassis, 2002). In 1994, Chan, Rastelli, and Pirrotta reported that a 2.2-kb PRE from *Ubx* caused silencing of the mini-*white* gene in the *pCaSpeR* vector. In that case, an *Ubx* PRE repressed mini-*white* expression even in the heterozygous state; many *Ubx* PRE-mini-*white* transformants had variegated eyes, some with the eye color only present in a few ommatidia. Using a different vector that contained both mini-*white* and *hs-neo* genes and selecting flies using G418 resistance, many of the lines obtained showed no detectable eye color and thus would not have been obtained if eye color had been used as the selectable marker. The 2.2-kb *Ubx bxd* PRE fragment they used was subsequently dissected into subfragments, six of which, when multimerized, mediated mini-*white* suppression (Horard, Tatout, Poux, & Pirrotta, 2000) in slightly different ways. Some subfragments gave transgenic flies with variegated eyes; one gave flies with patterned eyes (eye color in only part of the eye, but reproducibly in the same part of each eye). Of three multimerized subfragments tested for PREs in a maintenance assay in embryos, only one acted as a PRE. The data suggest that the 2.2-kb *Ubx bxd* PRE fragment contains multiple elements that function together to regulate *Ubx* expression throughout development.

Use of the FRT-Flp and LoxP-Cre recombinase systems to remove PREs from *pCaSpeR*-containing transgenes has shown that all PREs tested repress eye color to some extent in the heterozygous state. Figure 3.2 shows an example of this. In *P[PREmw]*, the 181-bp *en* PRE is flanked by LoxP sites. At this chromosomal insertion site, *P[PREmw]* flies have a light orange eye color. Removal of the PRE via Cre recombinase yields flies with a slightly darker eye color. The effect of the 181-bp *en* PRE is much more dramatic in the homozygous flies; with the PRE, the eye color is white; without the PRE, the eye color is red (Noyes et al., 2011).

Silencing of mini-*white* expression either in the heterozygous state or via PSS is the most common assay used to test the function of a presumptive PRE (Table 3.1). Is this a legitimate assay for PRE activity? There are a few reported cases where PREs that work in maintenance assays in embryos or larvae do not repress mini-*white* expression and vice versa (reviewed in Kassis, 2002), but, in general, there is an excellent correlation between PRE activity and PSS. Some minimal PREs may be stage or tissue specific and may not work in both assays. Results from our lab show a perfect correlation between fragments of *en/inv* DNA that mediate PcG repression in embryos and PSS (Americo et al., 2002; Brown, Grau, DeVido, & Kassis, 2005; Cunningham et al., 2010; DeVido et al., 2008). Thus, PSS is a good indication of PRE activity.

2.3. Effect of flanking regulatory DNA on PRE activity

In vivo, PREs can be located tens of kilobases away from the promoter they regulate. How are they able to act over such large distances? *In vitro*, PRC1 inhibits chromatin remodeling and transcription (reviewed in Simon & Kingston, 2009). Interestingly, PRC1 bound to a polynucleosome template *in vitro* can recruit a second template and inhibit chromatin remodeling and transcription of that template as well (Lavigne, Francis, King, & Kingston, 2004). Further, the *Drosophila* PRC1 subunit Psc can bind to and compact chromatin, an activity conserved in the M33 Polycomb homolog in mice (Grau et al., 2011).

The PSS assay suggests that PREs act together to facilitate gene silencing. The physical interaction between the PREs and promoters of the BX-C has been well documented by chromatin chromosome capture (3C) and fluorescent *in situ* hybridization (Lanzuolo, Roure, Dekker, Bantignes, & Orlando, 2007). Genome-wide chromatin conformation studies suggest that PcG proteins are a major organizer of the three-dimensional structure of the genome (recently reviewed in Delest, Sexton, & Cavalli, 2012; Pirrotta & Li, 2012). PcG proteins are present in "Polycomb bodies" where concentrations of PcG proteins appear as "dots" in the cell. Interestingly, PcG target genes colocalize with PcG bodies only in those cells where they are repressed (see Delest et al., 2012; Pirrotta & Li, 2012).

In a PRE-transgene assay, a PRE is removed from its normal chromatin context, often combined with regulatory DNA from a gene it normally does not regulate, and then inserted into a region of the genome where it normally does not reside. Given that PREs are designed to interact with other regulatory DNA, it is not surprising that the activity of a PRE in a transgene is dependent on where it is inserted in the genome. For example, PSS mediated by *en* PREs occurs in only about 50% of transgenic lines, and the extent of silencing also depends on the chromosomal insertion site (Americo et al., 2002; Noyes et al., 2011). Various models can be proposed to explain this. First, since PREs often work together, perhaps PSS only occurs in regions of the genome where there is another PRE nearby; there is some evidence to suggest that PRE-containing transgenes have a tendency to insert in the genome near PcG-regulated genes (reviewed in Kassis, 2002 and see below). Second, if the transgene inserts downstream of a promoter, it might be transcribed; transcription through a PRE has been shown to inactivate it (Schmitt, Prestel, & Paro, 2005). Finally, genomic enhancers flanking the PRE-transgene insertion site might activate mini-*white* expression via a mechanism not subject to PcG repression. A mammalian enhancer that can evict PcG proteins has recently been described (Vernimmen et al., 2011).

Recent work from our lab shows the effect of chromatin context (or flanking DNA) on PRE function. We carried out a genetic screen for dominant suppressors of PSS of an *en* PRE-mini-*white* transgene (*P*[*enPRE-mw*]), hoping to recover mutations in PcG genes or other genes necessary for the PSS activity of the *en* PRE (Noyes et al., 2011). We screened for suppressors of mini-*white* silencing of a *P*[*enPRE-mw*] transgene inserted on chromosome 2R between the genes *CG30456* and *GstS1*. We reasoned that if the mutations we recovered affected the activity of the *en* PRE directly, then they should suppress mini-*white* silencing by *P*[*enPRE-mw*] regardless of where it was inserted in the genome. Mutations that suppressed the mini-*white* silencing of a specific *P*[*enPRE-mw*] insertion must act on flanking regulatory DNA. Interestingly, most of the mutations we recovered suppressed the mini-*white* silencing at only one or a subset of *P*[*enPRE-mw*] insertion sites, again showing the profound effect of genomic location on PRE activity.

We have characterized two of the dominant suppressors of PSS by *P*[*enPRE-mw*] (Cunningham et al., 2012; Noyes et al., 2011). One was a dominant mutation in the gene encoding the transcriptional activator Without children (Woc^D) (Noyes et al., 2011). Woc^D is caused by a single amino acid change that we hypothesize increases the ability of Woc to activate transcription. Woc^D suppresses *P*[*enPRE-mw*]-mediated PSS at only two chromosomal insertion sites (out of 16 tested). This suggests that Woc^D does not act at the PRE directly but rather acts on flanking genomic DNA. Woc^D

may change the local chromatin conformation, perhaps by increasing the transcription of genes flanking the P[*enPRE-mw*] insertion site, inactivating PRE function at that chromosomal location.

The other PSS suppressor mutation we characterized was a dominant mutation in the cohesin-associated regulatory protein *wings apart-like* (*wapl*) (Cunningham et al., 2012), *wapl*^{AG}. Interestingly, *wapl*^{AG} mutants have a classic PcG phenotype, sex comb teeth on the second and third legs (Kennison, 1995), suggesting that Wapl is also a regulator of the homeotic gene *Scr*. Wapl acts in a protein complex with Pds5 to remove cohesin from chromosomes (Shintomi & Hirano, 2009). Our results suggest that increasing cohesin-binding stability antagonizes PcG silencing (Cunningham et al., 2012). The relationship between cohesin binding and PcG activity has been recently reviewed (Dorsett, 2011).

2.4. PREs can facilitate transcriptional activation or repression

The patterned expression of the *HOX* genes in *Drosophila* is initially set up by the gap and pair-rule transcription factors that directly bind HOX-regulatory DNA. Soon after, these direct regulators disappear, and the PcG and trithorax-group (trxG) proteins take over to maintain the repressed or active state, respectively. It is well established that PcG proteins act through PREs, and there is evidence that trxG proteins act through the same or closely linked DNA elements (trxG response elements, TREs, reviewed in Ringrose & Paro, 2004). Interestingly, in genome-wide studies, the C-terminus of Trithorax localizes to the same locations as PcG proteins, the presumptive PREs (Schuettengruber et al., 2009; Schwartz et al., 2010).

The first indication that PREs could maintain either the repressed or active chromatin state was for a DNA fragment from the BX-C, *Fab-7* (Cavalli & Paro, 1998). Remarkably, *Fab-7* was able to maintain either transcriptional repression or activation of two reporter genes throughout both mitosis and meiosis depending on the transcriptional state of the transgenes in the embryo. The authors designed a vector with a GAL4-inducible *lacZ* gene followed by the mini-*white* gene. Inclusion of the *Fab-7* fragment (that contains a PRE) repressed the expression of both *lacZ* and mini-*white*. After transient activation of the *lacZ* gene by GAL4 in embryos, both *lacZ* and mini-*white* expression were maintained throughout development. This fragment of DNA was subsequently called a "cellular memory module" or CMM. Three PRE-containing fragments from the BX-C have been shown to act as CMMs in this assay: *Fab7*, *Mcp*, *bxd*, and also a PRE from the *hedgehog* (*hh*) gene (Maurange & Paro, 2002; Rank et al., 2002). The *en* PRE did not act as a CMM in this assay (DeVido et al., 2008). Further experiments indicated that transcription through a PRE could turn it from a PRE to a TRE (Schmitt et al., 2005). One model states that it is the transcriptional state of PREs that determines whether they will maintain the active or repressed chromatin state. This may be true for some genes, but it is unlikely that this simple model will explain how the activity state is set for all PREs.

The ability of two closely linked *engrailed* PREs to mediate either transcriptional activation or repression was revealed in an enhancer-trap assay. An *en-lacZ* transgene containing 2.4 kb of upstream *en* upstream sequences (that includes the two PREs) driving *lacZ* expression had no intrinsic patterning activity on its own, but *lacZ* was expressed in patterns driven by genomic enhancers near the transgene insertion site. By flanking the *en* PREs with loxP or FRT sites, the expression of the *en-lacZ* transgene with and without the PREs in the same location could be examined. Remarkably, each PRE could act with flanking regulatory DNA as either a positive or negative regulatory element depending on the genomic context and the tissue examined. These data suggest that PREs, or closely associated sequences, mediate long-range interactions with enhancers or silencers around the PRE-transgene insertion site.

An excellent example of a fragment of DNA that acts as both a PRE and a presumptive TRE is provided by studies on a PRE from the *eve* locus (Fujioka et al., 2008). A 300-bp *eve* DNA fragment was shown to (1) mediate PSS of mini-*white*; (2) act as a PRE in an *Ubx*-reporter construct in embryos; and (3) maintain the continued expression of an *eve-lacZ* reporter construct in cells of the larval CNS; that is, without the PRE, the *eve-lacZ* gene was not expressed in the larval CNS. Remarkably, all three activities were dependent on a single binding site for the PcG-DNA binding protein Pho (Fujioka et al., 2008). Pho has been shown to interact with both PcG proteins and the trxG proteins Brahma, Moira, and Osa in *Drosophila* embryo extracts (Mohd-Sarip, Venturini, Chalkley, & Verrijzer, 2002). It would be interesting to know whether PREs from other genes could maintain the expression of the *eve-lacZ* reporter in the larval CNS, or whether this activity is specific for the *eve* PRE.

2.5. DNA binding proteins and sequence motifs associated with PREs

PREs are made up of multiple binding sites for many different factors. Despite the overwhelming amount of literature investigating the nature of *Drosophila* PREs, we still do not have a clear understanding of what precise combination of DNA binding sites/proteins are required for PRE activity. Studies of a number of different *Drosophila* PREs from *en* (Americo et al., 2002; DeVido et al., 2008; Kassis, 1994), *inv* (Cunningham et al., 2010), *eve* (Fujioka et al., 2008), *Ph* (Bloyer, Cavalli, Brock, & Dura, 2003), *hh* (Chanas & Machat, 2005; Maurange & Paro, 2002), *vestigal* (*vg*) (Okulski et al., 2011), and PREs of the BX-C: *Fab7*, *Mcp*, *bxd* (also called PRE$_D$), and *iab-2* (Busturia et al., 2001, Hagstrom, Müller, & Schedl, 1997, Mishra et al., 2001, Strutt, Cavalli, & Paro, 1997) have identified several different factors/binding sites as being important for PRE function. These include Pho/Phol, Sp1/KLF factors, GAGA/Psq, Zeste/Fs(1)h, Dsp1, Grainyhead (Grh), and other as yet unidentified factors (reviewed in Müller & Kassis, 2006; Ringrose & Paro, 2007) that we will discuss below.

Genome-wide ChIP studies are yielding specific information about the distribution of these proteins at PREs in different cell types and at different developmental stages (Kwong et al., 2008; Négre et al., 2006; Schuettengruber et al., 2009; Schwartz et al., 2006, 2010; Tolhuis et al., 2006). These studies show that the identified PRE-associated DNA binding proteins are not present at all PREs and bind to many sites that are not within PREs. The phenotypes of mutants of some of these DNA binding factors argue for a role of some of these proteins in *trx*G gene activation rather than PcG repression. This may reflect the close association of PREs and TREs.

In addition, genome-wide studies have shown that PcG repression is not the all or nothing system that was originally thought and is in fact much more dynamic. A study of PcG distribution in different tissue culture cells reflecting different activity states of some genes shows that PcG proteins can be bound to PREs in both the on and the off transcriptional state raising interesting mechanistic questions (Schwartz et al., 2010). Interestingly, Papp and Müller (2006) found binding of PcG proteins in both the on and off transcriptional states at *Ubx* PREs in imaginal disks. Comparison of embryo versus imaginal tissue PcG protein binding suggests the presence of stage-specific PREs (Kwong et al., 2008; Schuettengruber et al., 2009). The general consensus seems to be that no one factor alone is responsible for PRE activity and that PcG recruitment to PREs involves the cooperative contribution of a number of different DNA binding factors. The exact combination of factors that constitute a PRE may be different at different genes.

2.5.1 Pho and Phol

The only known bonafide member of the *PcG* genes that encodes a DNA binding protein is *pho*. *pho* zygotic mutants die as pharate adults with sex combs on the second and third legs (the "classic" PcG phenotype), while embryos derived from germ line clones that lack both maternal and zygotic *pho* die at the embryonic stage with segmentation defects (Breen & Duncan, 1986). The *Drosophila* genome encodes a protein highly related to Pho in the DNA binding domain, Pho-like (Phol) (Brown, Fritsch, Müller, & Kassis, 2003). *phol* mutants are homozygous viable with no homeotic phenotypes. Genetic experiments suggest that Pho and Phol act together to repress homeotic gene expression since *pho; phol* double mutants have more derepression of the homeotic gene *Ubx* in wing disks than do *pho* mutants. Phol also works independently of Pho; eggs derived from Phol females are fertilized but do not develop, a phenotype not observed for eggs that lack Pho (Breen & Duncan, 1986).

Pho and Phol are the *Drosophila* homologs of the mammalian proteins YY1 and YY2, respectively (Brown et al., 1998, 2003 Drews, Klar, Dame, & Bräuer, 2009; Nguyen, Zhang, Olashaw, & Seto, 2004). YY1 is a very dynamic protein that can be involved with repression, activation, and transcriptional initiation in a context-dependent way (Gordon, Akopyan, Garban, & Bonavida, 2006; Wang, Chen, & Yang, 2006). Pho has 96% amino acid identity with YY1 over four zinc fingers that are used for DNA binding and protein–protein interactions. There is also conservation of another region of the protein (18/22 amino acid identity between YY1 and Pho), a small spacer region implicated in protein–protein interactions. Phol has 80% amino acid identity to Pho over the four zinc fingers, and 10/22 and 11/22 amino acid identity to YY1 and Pho, respectively, in the spacer region. YY1 has been postulated to be involved in PcG repression in vertebrates (discussed in more detail below), and mammalian YY1 can partially rescue a *pho* zygotic mutant (Atchison, Ghias, Wilkinson, Bonini, & Atchinson, 2003).

Although Pho and Phol both bind to *Ubx* PRE$_D$ and Phol can silence *Ubx* transcription in the absence of Pho (Wang et al., 2004), genome-wide studies show that, in wild-type embryos, the distribution of Pho and Phol is different (Schuettengruber et al., 2009). Consistent with the PcG phenotype of zygotic *pho* mutants, Pho binding is highly correlated with the binding of Ph and Pc. Of genomic positions that share Ph and Pc binding, 96% correspond to Pho peaks but only 21% correspond to Phol peaks (Schuettengruber et al., 2009). The data suggest that Phol preferentially binds to a slightly different sequence than Pho and may be involved in transcriptional activation rather than repression (Schuettengruber et al., 2009). The 96% overlap between regions that bind Pc/Ph and Pho suggest that Pho could be absolutely required for PRC1 recruitment. However, in another study, also in embryos, there was only a 50% overlap of Pc binding sites with

Pho (Kwong et al., 2008). It should also be pointed out that there are many Pho sites that do not overlap with either Pc or Ph binding.

What does Pho do at the PRE? In embryos, Pho was found in two protein complexes, Pho-dINO80, a nucleosome-remodeling complex, and PhoRC where it is in a complex with dSfmbt, a protein that has methyl-lysine-binding activity (Grimm et al., 2009; Klymenko et al., 2006). Both Pho and Phol form stable complexes with dSfmbt in Sf9 cells. In contrast, Phol is not found associated with dINO80. dSfmbt is also bound to PREs, and mutations in the gene encoding *dSfmbt* cause homeotic derepression (Klymenko et al., 2006; Oktaba et al., 2008) showing it is also important for PcG repression. Therefore, one function of Pho and Phol may be to bring dSfmbt to the PRE. Biochemically, Pho has been reported to interact with components of both PRC1 and PRC2 in coimmunoprecipitation and GST pulldown assays (Mohd-Sarip, Cleard, Mishra, Karch, & Verrijzer, 2005; Wang et al., 2004). Further, both Pho and dSfmbt copurified with a Pc protein labeled *in vivo* by biotinylation (Strübbe et al., 2011). Pho and Phol are thought to play a key role in the binding of both PRC1 and PRC2 to chromatin (Wang et al., 2004). Interestingly, the PcG protein Scm can bind to an *Ubx* PRE in the absence of Pho/Phol (Wang et al., 2010).

There is no doubt that Pho plays a key role at PREs. To our knowledge, all PREs tested in either the mini-*white* silencing or the embryonic or larval maintenance assays bind Pho. Mutation of Pho binding sites within PREs abrogated the function of all PREs where it has been tested (reviewed in Kassis, 2002; Fujioka et al., 2008). There is some data that other PRE-binding proteins may be necessary to facilitate the binding of Pho to the PRE. For example, Grh, GAF, and Dsp1 have all been postulated to cooperatively assist Pho binding to chromatinized substrates (Blastyák, Mishra, Karch, & Gyurkovics, 2006; Dejardin et al., 2005; Mahmoudi, Zuijderduijn, Mohd-Sarip, & Verrijzer, 2003). Pho is generally thought to have a repressive role but, as discussed above, Fujioka et al. (2008) have shown that Pho can act as an activator in a specific subset of cells in the *Drosophila* nervous system. It remains to be seen whether this activating role of Pho is specific for this case or can be generalized to many PRE/TREs.

2.5.2 Sp/KLF proteins

An analysis of a PRE from the *en* gene identified a Sp1/KLF binding site as being required for PSS and PRE activity of a 181-bp PRE (Americo et al., 2002; Brown et al., 2005). There are nine members of the Sp1/KLF family of proteins in *Drosophila* and *in vitro* transcription/translation of the DNA binding domains showed that 8/9 of the proteins were capable of binding to the site in the *en* PRE (Brown et al., 2005). As seen in the extensive family of mammalian Sp1/KLF proteins (reviewed in Kaczynski, Cook, & Urrutia, 2003), there were differences in the binding site preferences of the *Drosophila* family members. Further analysis identified a member of this group, Spps (Sp1 factor for PSS) as being involved in PRE activity of the *en* PRE (Brown & Kassis, 2010). Flies mutant for *Spps* die as late pharate adults and *Spps* enhances the *pho* mutant phenotype. This indicates that Pho and Spps either work together or in parallel parts of the same pathway to silence PcG target genes. PSS of mini-*white* in constructs carrying either an *en* PRE or a *bxd* PRE is lost in *Spps* mutants, showing that Spps is required for PSS. ChIP-qPCR showed that Spps binds to both the *en* and *bxd* PREs in larvae. ChIP experiments showing the genome-wide distribution of Spps have not yet been published. However, a general role for Spps in PRE activity is predicted given that Spps colocalizes almost perfectly with Psc in polytene chromosomes (Brown & Kassis, 2010). Mammalian YY1 and Sp1 have been shown to be able to interact (Lee, Calvin, & Shi, 1993; Seto, Lewis, & Shenk, 1993) which raises the possibility for interaction here. The involvement of other members of the *Drosophila* Sp1/KLF factors in PcG repression has yet to be determined. The mammalian Sp1/KLF proteins are a dynamic set of proteins involved in many aspects of gene expression, differentiation, and cancer (Kaczynski et al., 2003; McConnell & Yang, 2010).

2.5.3 GAGAG sites

The sequence GAGAG is required for the activity of many PREs (reviewed in Kassis, 2002; Fujioka et al., 2008). GAGAG elements are binding sites for two proteins, GAGA factor (GAF) and Pipsqueak (Psq) (Lehmann, Siegmund, Lintermann, & Korge, 1998). GAF recruits a chromatin-remodeling complex and the role of GAF in PRE activity may be to remove nucleosomes at the PRE, allowing the binding of other transcription factors (Mahmoudi et al., 2003). Genome-wide studies show that GAF is present at about 50% of sites bound by both Pc and Ph (Négre et al., 2006; Schuettengruber et al., 2009). Psq also binds to the GAGAG sequence although it is reported to have a preference for longer GAGA sequences in keeping with the fact that it has a different DNA binding domain (HLH; Siegmund & Lehmann, 2002) from GAF (a single zinc finger; Lehmann et al., 1998). Psq has been isolated from S2 cells in a protein complex (CRASCH; Huang & Chang, 2004; Huang, Chang, Yang, Pan, & King, 2002) with PcG proteins. The genome-wide distribution of Psq binding is not known but GAF and Psq proteins colocalize on polytene chromosomes. GAF and Psq both have the same type of protein–protein interaction domains, BTB domain, and can interact with each other (Schwendemann & Lehman, 2002) and with some other members of this class of proteins, for example, Tramtrack (Pagans, Ortiz-Lombardia, Esinas, Bernues, & Azorin, 2002). GAF is not a stable component of any of the PcG complexes isolated to date. Recently, a vertebrate homolog of GAF was identified that also binds to GAGAG sequences and shows extensive homology with GAF (Matharu, Hussain, Sankaranarayanan, & Mishra, 2010). It will be interesting to know whether this protein plays a role in PcG silencing in mammals.

GAF is encoded by the gene *trithorax-like* (*trl*; Farkas et al., 1994). As suggested by the name, *trl* mutants have a trxG phenotype. GAF is required for many aspects of gene expression and also for nuclear division in *Drosophila* embryos (Bhat et al., 1996). GAF is bound to GAGA satellite sequences. Because of its multifunctional role in controlling gene expression and cell division, it is impossible to directly test the role of GAF at PREs in embryos. Psq acts during oogenesis in the establishment of the anterior/posterior axis of the oocyte (Siegel, Jongens, Jan, & Jan, 1993). A role for Psq in PcG silencing is suggested by a genetic interaction between a *psq* and *Pc*. A mutation in *psq* strongly enhances the derepression of *Ubx* in wing and leg disks of larvae heterozygous for a *Pc* allele, strongly suggesting Psq plays a role in PcG silencing (Huang et al., 2002). The exact role of GAF versus Psq at the PREs needs further exploration.

2.5.4 GTGT sequence

This sequence has shown up in a number of studies to be enriched in PRE sequences (Ringrose et al., 2003; Schuettengruber et al., 2009). Deletion of the GTGT sequences in the *vg* PRE led to reduction in silencing activity of this fragment suggesting that this sequence may play a role in PcG repression (Okulski et al., 2011). Interestingly, in mammals, the Sp1/KLF factor Sp2 was shown to bind to a GTGT box (GGTGTGGGG) implicated in regulating T-cell receptor gene expression (Kaczynski et al., 2003; Kingsley & Winoto, 1992; Suske, Bruford, & Philipsen, 2005). It would be interesting to know whether the GTGT sequences deleted in the *vg* PRE bind to Spps or other *Drosophila* Sp1/KLF family members.

2.5.5 Dsp1

Dsp1 was identified as a PRE DNA binding protein important for the recruitment of PcG complexes to polytene chromosomes (Dejardin et al., 2005). These authors reported that Dsp1 binds to the consensus sequence GAAAA and that mutation of GAAAA within the *en* and *Fab7* PREs not only abrogated PSS of mini-*white* but also changed the PREs into constitute TREs. These data suggest that Dsp1 is required for PRE repression activity.

Genome-wide studies with 4–12 h embryo extracts have shown that Dsp1 binds to a subset of PREs (about 50% of the Ph/Pc sites identified (225/439); Schuettengruber et al., 2009). Curiously, the GAAAA consensus sequence is not enriched within these PREs. Clearly, the DNA binding site of Dsp1 requires further definition; however, Dsp1 might recognize DNA structure instead of sequence. Dsp1 is a member of the class of HMG proteins that have two HMG domains; these proteins bind to the minor groove of DNA without sequence specificity, instead recognizing DNA structural features. This type of HMG protein has been reported to be able to introduce a significant distortion or bend into the DNA (for reviews, see Agresti & Bianchi, 2003; Stros, 2010). Dsp1 could facilitate PcG protein binding or long-range interactions by changing the structure of the DNA.

Mutants of Dsp1 exhibit features more typical of a *trxG* gene than a *PcG* gene (Decoville, Giacomello, Leng, & Locker, 2001; Rappailles, Decoville, & Locker, 2005). Dsp1 has been shown to coimmunoprecipitate with the chromo domain containing protein, Corto (Salvaing et al., 2006). These authors suggest that Dsp1 binds to the *Scr* PRE only when *Scr* is being transcribed, thus Dsp1 plays a critical role in TRE activity. Clearly, more work needs to be done to understand the role of Dsp1 in PcG and trxG activity.

2.5.6 Grh

Grh was found to bind to the *iab-7* PRE and to cooperatively interact with Pho both *in vitro* and genetically (Blastyák et al., 2006). The binding site for Grh identified in the BX-C *iab-7* PRE was reported to be TGTTTTTT. However, other groups report a Grh consensus sequence as WCHGGTT (where W is A or T and H is not G) (Almeida & Bray, 2005; Venkatesan, McManus, Mello, Smith, & Hansen, 2003). To date, a genome-wide study of Grh binding in relationship to PcG protein binding has not been reported. A presumptive Grh binding site is present in an *eve* PRE (Fujioka et al., 2008) but not in the 181-bp *en* PRE (J. Lesley Brown and Judith A. Kassis, unpublished observation), suggesting it may be important for only a subset of PREs.

Other evidence that Grh may be involved in PcG repression comes from work on the mammalian Grh-family member CP2. CP2 was shown to interact with a mammalian Ring protein, DinG. The CP2–DinG interaction was necessary for transcriptional repression (Tuckfield et al., 2002). The authors also showed that *Drosophila* Grh could interact with dRing in Gst-pull down experiments *in vitro*.

Drosophila Grh regulates the expression of many genes and can act as either a transcriptional repressor or activator dependent on the context (see references cited in McQuilton, Pierre, Thurmon, & The FlyBase Consortium, 2012). Of particular interest for this review, Grh, Zeste, and GAGA binding sites activate transcription of an *Ubx* promoter reporter construct through intermingled clusters of binding sites (Hur, Laney, Jeon, Ali, & Biggin, 2002 and references therein). Interestingly, Hur et al. (2002) found that Zeste and GAGA promoter-proximal binding sites were able to maintain PcG-mediated repression of a 22-kb *Ubx-lacZ* reporter construct in embryos, whereas Grh binding sites did not have this activity. The role of Grh in PcG repression requires further study.

2.5.7 Zeste binding sites

The role of Zeste in PcG repression is also unclear. Zeste has been reported to be an activator or repressor of transcription dependent on the context and has been reported to be a stoichiometric component of PRC1 (Saurin et al., 2001). Schuettengruber et al. (2009) found only a 25% overlap of Ph with Zeste in genome-wide binding data. Oktaba et al. (2008) report very little overlap between Zeste and Pho in genome-wide binding data. Zeste mutants are homozygous viable and fertile and do not show PcG phenotypes (Goldberg, Colvin, & Mellin, 1989). *female sterile (1) homeotic (fs(1)h)* is a member of the *trxG* genes. *fs(1)h* encodes a double bromodomain-containing protein that binds to Zeste sites within the *Ubx* promoter and activates transcription of the *Ubx* gene (Chang, King, Lin, Kennison, & Huang, 2007). It is not known if Fs(1)h binds to Zeste sites in PREs.

2.6. Are all PREs alike?

Algorithms based on clustered consensus binding sites for PRE DNA binding proteins have been marginally successful in identifying PREs, giving both false positive and false negative results (Fiedler & Rehmsmeier, 2006; Ringrose et al., 2003; Zeng, Kirk, Gou, Wang, & Ma, 2012). In one study, of 167 potential PREs that Ringrose et al. (2003) identified, only 16% of them were PcG binding sites in PcG ChIP genome-wide studies of *Drosophila* embryos

(Schuettengruber et al., 2009). A weakness of this approach is the unreliability of the consensus binding sequences that are used in the search. A consensus sequence is our best guess but it does not take into account cooperativity between weak sites, the influence of nearby binding proteins, or the fact that a protein may be able to bind to sites that deviate from the consensus. For example, in a study on PREs at the *invected* locus (Cunningham et al., 2010), a fragment that had both PSS and PRE activity did not contain any matches to the core consensus binding sequence for Pho (GCCAT); thus this fragment was not predicted to be a PRE by the jPREdictor algorithm; however, this fragment clearly bound Pc and Ph in ChIP studies, suggesting it was a PRE. There were several CCAT sequences in this PRE, and 5/6 bound Pho *in vitro*. In addition, some PREs require Pho binding sites that contain sequences beyond the minimal Pho consensus. Two extended Pho binding sites were required to interact and recruit Pc complexes to chromatin in the *iab-7* PRE (Mohd-Sarip et al., 2002; reported to be a PcG target Mohd-Sarip et al., 2005, 2006). In studies examining the genome-wide distribution of Pho, Oktaba et al. (2008) noted that an extended Pho consensus sequence was enriched in regions that bound both Pho and dSfmbt. They also examined the number of Pho consensus sites within a Pho-bound region and found that, while often there was more than one Pho site, the spacing of these sites was variable, and many fragments had only one Pho binding site (Oktaba et al., 2008). Further, the number and order of Pho and other motifs varies greatly between orthologous PREs in different *Drosophila* species (Hauenschild, Ringrose, Altmutter, Paro, & Rehmsmeier, 2008); clearly there is great flexibility in PRE-binding site configuration.

PRE prediction algorithms make the assumption that all PREs have the same requirements for activity. This seems unlikely given the complexity of the different PcG target loci and the indication that different PREs function at different times in development. There is also a difference in the arrangement of PREs within a gene that may reflect underlying differences in the way the PRE works. A subset of PREs are located immediately upstream of the promoter, while others are many kilobases away. The available data suggest that some DNA binding proteins are present at most PREs (Pho), for example; however, it is likely that locus specific factors influence the activity of at least a subset of PREs.

A limited number of PREs have been shown to work with regulatory DNA from other genes (discussed above). For example, PREs from the genes *en*, *eve*, and Abd-B maintain silenced expression of an *Ubx-lacZ* reporter construct in the correct pattern in embryos. Further, the *iab-7* and *iab-8* core PREs completely replaced the activity of a core *bxd* PRE in the endogenous *Ubx* gene (Kozma, Bender, & Sipos, 2008). In general, these studies used minimal PRE fragments of only a few hundred base pairs or less that contained binding sites for Pho and GAF. Larger PRE fragments are more likely to contain other regulatory DNA and may not work in these assays. For example, while a 181-bp minimal *en* PRE maintained an *Ubx*-like expression pattern of an *Ubx-lacZ* transgene in embryos, a larger fragment (2.6 kb) that contained the 181-bp PRE as well as an additional PRE and flanking DNA totally silenced the *Ubx-lacZ* reporter gene in a *Pc*-dependent way (Americo et al., 2002). Recently, Okulski et al. (2011) used the phi-C31 system to integrate two different PRE-mini-*white* transgenes (*vg* and *Fab7*) at four different chromosomal insertion sites and quantitated the amount of mini-*white* repression by the PREs both in flies heterozygous and homozygous for the transgene. They noted that the *vg* and *Fab7* PREs gave quantitatively different amounts of mini-*white* silencing and behaved differently at the different insertions sites. Although this was a very careful study, one problem with it was that the PRE fragments they used were 1.6 kb and were likely to have other regulatory DNA for their respective loci. In this regard, the 181-bp *en* PRE also has promoter-tethering activity (Kwon et al., 2009). It is important to keep in mind that PREs act in the context of other regulatory DNA and that other regulatory elements may be closely linked or even intertwined.

Strübbe et al. (2011) recently purified Pc under physiological conditions designed to maintain the weakest associations and then analyzed copurifying products by mass spectrometry. This identified all of the PRC1 core components Pc, Psc, Sce/dRing, Ph and Su(z)2 and PhoRC (dSfmbt and Pho) but components of PRC2 were not found. Fs(1)h and Grh also copurified. In addition to known PcG complexes, the data also showed unexpected interactions with the cohesin complex and with proteins related to the Moz/Morf histone acetyltransferase complex. This type of approach, combined with a rigorous analysis of individual PREs will be necessary to understand how PcG proteins are recruited.

3. CHARACTERISTICS OF PcG TARGET GENES IN *DROSOPHILA*

3.1. Many PcG targets have multiple PREs

PcG target genes tend to have a large amount of regulatory DNA reflecting their complex and often dynamic expression patterns throughout development (Kharchenko et al., 2011; Négre et al., 2006). PcG target genes often occur in gene clusters. The best known of these are the *HOX* gene clusters including the BX-C that contains the homeotic genes *Ubx*, *Abd-B*, and *Abd-A*. The arrangement and function of PREs within the BX-C have been extensively studied (reviewed in Maeda & Karch, 2006, 2011); in fact, much of what we know about PREs comes from studies on BX-C PREs. Here, we describe the PREs of three PcG target loci, *eve*, an example of a single gene PcG target and two PcG targets with two coregulated genes, *en*/*inv* and *Psc*/*Su(z)2*.

3.1.1 eve

The *eve* gene is a pair-rule gene required for segmentation and the development of specific lineages in the mesoderm and nervous system. A single *eve* 1.5-kb transcription unit is flanked by about 9 kb of regulatory DNA downstream and 6-kb upstream, containing many *eve* enhancers (Fujioka et al., 1999). *PcG* genes regulate *eve* expression in the nervous system (Fujioka et al., 2008; Smouse, Goodman, Mahowald, & Perrimon, 1988). In S2 cells, a 20-kb "*eve* domain" is covered with the characteristic H3K27me3 PcG mark generated by PRC2 (Kharchenko et al., 2011). There are two fragments of *eve* DNA that mediate PSS of mini-*white*, a 300-bp fragment located 8.9-kb downstream of the *eve* transcription unit (*eve* PRE300), and another 441-bp fragment that includes the *eve* promoter. *eve* PRE300 acts as a PRE in a *Ubx-lacZ* reporter gene and also as a positive element to maintain expression of an *eve*-reporter gene in specific cells of the larval nervous system (discussed above). The *eve* proximal-promoter PRE has not been tested in these assays. Both *eve* PREs contain Pho, GAF, and Zeste binding sites (Fujioka et al., 2008). Interestingly, *eve* PRE300 is not required for correct *eve* expression in an *eve* rescue transgene (Fujioka et al., 1999). This suggests that the promoter-proximal *eve* PRE is sufficient for maintaining *eve* expression.

3.1.2 en/inv

The *en* and *inv* genes are coexpressed genes that share regulatory DNA (Gustavson et al., 1996) and form a 115-kb H3K27me3 domain in S2 cells, embryos, and larvae (Kharchenko et al., 2011). There are four well-characterized PREs in the *inv*/*en* locus; two closely linked ones located between −400 bp and −2 kb upstream of the *en* transcription start site (DeVido et al., 2008) and two upstream of *inv*, one coincident with the major *inv* transcription start site and another about 6-kb upstream of that (Cunningham et al., 2010). In S2 cells and adults, PcG proteins are bound to a single 2-kb peak just upstream of the *en* transcription start site (encompassing the two PREs) and to the two *inv* PREs, however, in embryos and larvae, PcG proteins bind to additional regions upstream of the *en* gene (Négre et al., 2006; Oktaba et al., 2008). It is not known whether these other PcG binding regions represent additional stage-specific PREs or are the result of the three-dimensional structure of the locus causing cross-linking to fragments because of chromatin compaction. PRE activity assays will need to be done on these other fragments to distinguish between these two possibilities. Interestingly, deletion of a 1.5-kb fragment ($en^{\Delta 1.5}$, that includes the two PREs upstream of the *en* transcription start site) from the endogenous *en/inv* locus does not alter the pattern of *en* or *inv* expression (Cheng, Kwon, Arai, Mucci, & Kassis, 2012). These data suggest that PREs might act redundantly in the *en/inv* domain; however, it is not know whether deletion of the PREs in $en^{\Delta 1.5}$ causes quantitative difference in *en/inv* expression.

3.1.3 Psc/Su(z)2

The PcG group genes *Psc* and *Su(z)2* exist in an H3K27me3 domain that extends over about 91 kb in S2 cells; insulator elements may limit the spread of H3K27me3 into flanking genes (Park et al., 2012). *Psc* and *Su(z)2* are ubiquitously expressed genes, so, obviously, PcG proteins do not silence gene expression in this case. It is thought that PcG proteins regulate the level of *Psc* and *Su(z)2* expression. ChIP experiments in S2 cells show four large discrete Psc binding sites as well as additional smaller peaks (Schwartz et al., 2006). Park et al. (2012) showed that two of these peaks had PRE activity in both a mini-*white* silencing assay and in a larval *Ubx*-reporter gene functional assay. Another Psc binding fragment behaved as a downregulatory module rather than a typical PRE. Deletion of one of the *Psc/Su(z)2* PREs from the endogenous locus caused a twofold increase in the level of *Psc* and *Su(z)2* transcription suggesting that this element downregulates the level of transcription. Why do PREs from the *Psc/Su(z)2* gene act as silencers in transgenes, but not silence expression of the endogenous gene? One hypothesis is that the type of enhancers associated with this gene are not subject to PcG silencing. More work needs to be done on this interesting locus.

3.2. Homing and PcG target genes

P-element based vectors have been used to make transgenic *Drosophila* for about 30 years (Rubin & Spradling, 1982). Aside from some hotspots and a preference to insert near the $5'$ end of transcription units, P-based vectors insert in the genome in a relatively nonselective manner and as such, have been used as a tool to generate thousands of transposon-induced mutations in *Drosophila* (Bellen et al., 2011). In 1990, Hama, Ali, and Kornberg reported that P-based constructs that contain *en* regulatory DNA (*P[en]*) insert near or into the *en/inv* domain at a very high frequency. They called this phenomenon "homing" (Hama, Ali, & Kornberg, 1990). Further studies on *P[en]* homing by *en* regulatory DNA showed that a 2-kb fragment, extending from −2.4- to −0.4-kb upstream of the *en* transcription start site was sufficient for homing, and that the *en* PREs contributed to the homing activity of this fragment (Cheng et al., 2012). Further, *P[en]* inserted near PcG targets at an increased frequency compared with a P-based vector used to generate insertions for the *Drosophila* gene disruption project (Cheng et al., 2012). These data suggest that either the chromatin structure or PcG proteins themselves bring *P[en]* to the region of PcG-regulated loci. This is consistent with the view that PcG targets colocalize in the nucleus in PcG bodies (Delest et al., 2012; Pirrotta & Li, 2012) and suggest that these bodies also occur in germ cells where homing occurs. Transgenes containing *polyhomeotic* PREs and BX-C PREs have also been reported to insert at a high frequency in or near PcG target genes (reviewed in Kassis, 2002).

Fragments of DNA from the BX-C and *eve* PcG targets have been shown to mediate homing (Bender & Hudson, 2000; Fujioka, Wu, & Jaynes, 2009). Like *P[en]*, insertions mediated by the BX-C and *eve* homing fragments occurred over a large region, causing insertions throughout and near the parent locus. For these two cases, the fragments that mediate homing are most likely insulators, not PREs (Bender & Hudson, 2000; Fujioka et al., 2009). Anecdotally, 1/5 insertions of a P-based vector that contained one of the *Psc/Su(z)2* PREs was inserted near the promoter of the *Su(z)12* gene, strongly suggesting that this fragment mediates homing. Note that one reported case of homing, that of the *linotte* gene (now called *drl*; Taillenbourg & Dura, 1999), is not reported to be a PcG target gene in embryos, larvae, or S2 cells; however, it is possible it could be a PcG-regulated gene in germ cells where P-element homing occurs.

4. PREs IN VERTEBRATES

PcG proteins in vertebrates are much more diverse and play a role in gene regulation in many different tissues and stages of development (two recent reviews: Beisel & Paro, 2011; Kerppola, 2009), including

X-inactivation (one recent review: Jeon, Sarma, & Lee, 2012). The mechanism of PcG protein recruitment varies dependent on the target and can be either RNA or protein mediated. In vertebrates, there are gene-specific DNA binding proteins that recruit specific PcG proteins to specific genes (reviewed in Beisel & Paro, 2011; Kerppola, 2009; Schuettengruber & Cavalli, 2009). Here, we focus on vertebrate DNA fragments that have been shown to have properties similar to *Drosophila* PREs.

In ES cells, CpG islands have been shown to be required for PRC2 recruitment (for review, see Deaton & Bird, 2011). In one elegant study, Mendenhall et al. (2010) showed that a CpG island was sufficient for recruitment of PRC2 to a BAC transgene integrated in ES cells. Binding sites for transcriptional activators in or next to the GC-rich sequence abrogated its ability to recruit PRC2 pointing to a competition between transcriptional activation and repression. Remarkably, even a GC-rich region from *E. coli* could recruit PRC2 when integrated into the transgene. Interestingly, PRC1 was not recruited to the DNA in their experiments. Lynch et al. (2012) provide a definitive PRE assay in mouse ES cells at the α-globin gene locus. They used recombination-mediated cassette exchange to study PcG protein binding and histone modifications to a single-copy human α-globin locus integrated into a mouse ES cell line. They also showed the importance of CpG islands in PcG recruitment and their experiments also pointed to a competition between transcriptional activation versus PcG repression. In addition, a key role for unmethylated CpG dinucleotides in PcG recruitment was established as they found *de novo* sites of PcG recruitment at CpG-rich sequences in methyltransferase-deficient ES cells. It is not known if PRC2 recruitment to CpG-rich sequences is protein or RNA mediated or occurs by default in the absence of transcriptional activators.

Are there any PREs in mammals that have binding sites for proteins important for PRE activity in *Drosophila*? Pho/Phol (YY1-homologues), Spps (a Sp1/KLF family member), Dsp1 (a HMG protein), and GAF all have mammalian homologues. In embryonic and neural stem cells, YY1 and PRC2 show almost no colocalization but, instead, YY1 binding is correlated with transcribed genes (Mendenhall et al., 2010). However, there is some genetic and biochemical evidence that YY1 might recruit PcG proteins at some loci (Garcia, Marcos-Gutierrez, del Mar Lorente, Moreno, & Vidal, 1999; Lorente et al., 2006). Below, we describe three examples of vertebrate PREs, studied in tissues other than ES cells that bear some similarities to *Drosophila* PREs.

Sing et al. (2009) identified a putative PRE (PRE-*kr*) that regulates expression of the *Maf/Kreisler* gene in rhomdomeres in mice. Endogenous PRE-*kr* binds PRC1 and PRC2 components. A reporter construct containing PRE-*kr* recruits PcG proteins. Like *Drosophila* PREs, PRE-*kr* contains binding sites for YY1 and GAF. The role of the YY1 and GAF binding sites in PRE-*kr* function has not been tested. Interestingly, PRE-*kr* mediates mini-*white* silencing and recruits PcG proteins to polytene chromosomes in *Drosophila* showing that this 3-kb fragment also acts as a PRE in *Drosophila*. Further studies would be necessary to determine if the same sequence motifs are required for PRE activity in the mouse and fly.

A vertebrate PRE from a *HOX* gene also shows characteristics of a *Drosophila* PRE. Woo, Kharchenko, Daheron, Park, and Kingston (2010) identified a 1.8-kb region between the human *HOXD11* and *HOXD12* genes (D11.12) that is bound by PcG proteins at the endogenous locus and functions as a PcG-dependent repressor in reporter constructs. Analysis of this element showed that there is a cluster of conserved YY1 sites important for the recruitment of BMI1 (a PRC 1 component) but less so for the recruitment of the PRC2 component SUZ12. A 237-bp region that is highly conserved across vertebrates was essential for the recruitment of members of both PRC1 and PRC2. The D11.12 region also contains a CpG island but the functional importance of this has not been tested. These results imply that PRC1 and PRC2 have different methods of recruitment.

A potential human PRE was identified in resting T cells downstream of the *SLCA17* gene (Cuddapah et al., 2012). In resting T cells, the *SLCA17* gene is not transcribed and PcG proteins are bound to the presumptive PRE. Treatment of HeLa cells with RNAi to SUZ12 caused a decrease in SUZ12 levels and to the levels of Bmi1 and Ring1B binding to the presumptive *SLCA17* PRE. A 3-kb putative *SLCA17* PRE contains YY1 and GAF binding sites and acted as a pairing-sensitive silencer in *Drosophila*, again suggesting that PcG proteins in *Drosophila* and vertebrates can be recruited by similar mechanisms (Cuddapah et al., 2012).

5. OUTLOOK

Much has been learned about *Drosophila* PREs since their discovery almost 20 years ago. First discovered as silencers of homeotic gene expression, genome-wide studies have revealed hundreds (or, in some studies, thousands) of potential PREs in *Drosophila*. Available data suggest that not all PREs are alike, but the differences between PREs are not well defined.

The situation in mammals is even more complex with the increased diversity of mammalian PcG proteins and the involvement of PcG proteins in so many diverse tissues, stages of development, and processes. It is still too early to conclude whether there is a class of mammalian PREs that closely resemble *Drosophila* PREs. There are many fundamental questions about *Drosophila* PREs that need to be addressed to understand PRE composition and function; some of the questions we have discussed in this review are summarized below.

(a) Are all PREs alike? Functional assays on hundreds of potential PREs using the phi-C31 integration system and reporter assays for different developmental stages will be necessary to assess whether there are different classes of PREs. In these assays, it will be important to use minimal PRE fragments (hundreds of base pairs) as larger DNA fragments will most likely have gene-specific activities including enhancers and silencers that could interfere with PRE assays.

(b) Mutations in different *PcG* genes in *Drosophila* (and mammals) lead to different phenotypes suggesting that not all the PcG proteins act together all the time or at all targets. Do different PREs recruit different PcG complexes or does the nature of the gene activators determine which PcG proteins are required at a particular target gene?

(c) At some targets, PcG proteins are bound to transcribed genes. What do PcG proteins do at actively transcribed genes? How can PcG proteins repress transcription at some targets and silence it at others?

(d) Many PcG target genes are expressed in many different tissues at many different times in development; only a subset of the expression patterns is subject to PcG regulation. Are there some types of enhancers (or promoters for that matter) that cannot be regulated by PcG proteins?

(e) How do PRE-binding proteins recruit or anchor PcG protein complexes? The mechanisms are far from known.

Despite all these questions, it is quite clear that PREs play a major role in regulating the expression of many genes in the *Drosophila* genome. It is incumbent on us to figure out how they work.

ACKNOWLEDGMENTS

We thank Yuzhong Cheng and Payal Ray for comments on this chapter. We thank the authors of many recent reviews on this topic; their comprehensive reviews made it easier for us to highlight only a subset of the literature on this important topic. We apologize for any omissions or misstatements of data. J. A. K. and J. L. B. are supported by the intramural research program of the National Institutes of Health and *Eunice Kennedy Shriver* National Institute of Child Health and Human Development.

REFERENCES

Agresti, A., & Bianchi, M. (2003). HMGβ proteins and gene expression. *Current Opinion in Genetics and Development, 13*, 170–178.

Almeida, M. S., & Bray, S. J. (2005). Regulation of post-embryonic neuroblasts by *Drosophila* Grainyhead. *Mechanisms of Development, 122*, 1282–1293.

Americo, J., Whiteley, M., Brown, J. L., Fujioka, M., Jaynes, J. B., & Kassis, J. A. (2002). A complex array of DNA-binding proteins required for pairing-sensitive silencing by a Polycomb group response element from the Drosophila *engrailed* gene. *Genetics, 160*, 1561–1571.

Atchison, L., Ghias, A., Wilkinson, F., Bonini, N., & Atchinson, M. L. (2003). Transcription factor YY1 functions as a PcG protein *in vivo*. *The EMBO Journal, 22*, 1347–1358.

Barges, S., Mihaly, J., Galloni, M., Hagstrom, K., Muller, M., Shanower, G., et al. (2000). The fab-8 boundary defines the distal limit of the Bithorax complex iab-7 domain and insulates iab-7 from initiation elements and a PRE in the adjacent iab-8 domain. *Development, 127*, 779–790.

Beisel, C., & Paro, R. (2011). Silencing chromatin: Comparing modes and mechanisms. *Nature Reviews Genetics, 12*, 123–135.

Bellen, H. J., Levis, R. W., He, Y., Carlson, J. W., Evans-Holm, M., Bae, E., et al. (2011). The Drosophila gene disruption project: Progress using transposons with distinctive site specificities. *Genetics, 188*, 731–743.

Bender, W., & Hudson, A. (2000). P element homing to the Drosophila bithorax complex. *Development, 127*, 3981–3992.

Bhat, K. M., Farkas, G., Karch, F., Gyurkovics, H., Gausz, J., & Schedl, P. (1996). The GAGA factor is required in the early *Drosophila* embryo not only for transcriptional regulation but also for nuclear division. *Development, 122*, 1113–1124.

Blastyák, A., Mishra, R. K., Karch, F., & Gyurkovics, H. (2006). Efficient and specific targeting of Polycomb group proteins requires cooperative interaction between Grainyhead and Pleiohomeotic. *Molecular and Cellular Biology, 26*, 1434–1444.

Bloyer, S., Cavalli, G., Brock, H., & Dura, J.-M. (2003). Identification and characterization of *polyhomeotic* PREs and TREs. *Developmental Biology, 261*, 426–442.

Breen, T. R., & Duncan, I. M. (1986). Maternal expression of genes that regulate the Bithorax complex of Drosophila melanogaster. *Developmental Biology, 118*, 442–456.

Brown, J. L., Fritsch, C., Müller, J., & Kassis, J. A. (2003). The Drosophila *pho-like* gene encodes a YY1-related DNA binding protein that is redundant with pleiohomeotic in homeotic gene silencing. *Development, 130*, 285–294.

Brown, J. L., Grau, D. J., DeVido, S. K., & Kassis, J. A. (2005). An Sp1/KLF binding site is important for the activity of a Polycomb group response element from the Drosophila *engrailed* gene. *Nucleic Acids Research, 33*, 5181–5189.

Brown, J. L., & Kassis, J. A. (2010). Spps, a Drosophila Sp1/KLF family member binds to PREs and is required for PRE activity late in development. *Development, 137*, 2597–2602.

Brown, J. L., Mucci, D., Whiteley, M., Dirksen, M. L., & Kassis, J. A. (1998). The *Drosophila* Polycomb group gene *pleiohomeotic* encodes a DNA binding protein with homology to the transcription factor YY1. *Molecular Cell, 1*, 1057–1064.

Busturia, A., Lloyd, A., Bejarano, F., Zavortink, M., Xin, H., & Sakonju, S. (2001). The MCP silencer of the Drosophila Abd-B gene requires both Pleiohomeotic and GAGA factor for the maintenance of repression. *Development, 128*, 2163–2173.

Cao, R., Wang, L., Wang, H., Xia, L., Erdjument-Bromage, H., Tempst, P., et al. (2002). Role of histone H3 lysine 27 methylation in Polycomb-group silencing. *Science, 298*, 1039–1043.

Cavalli, G., & Paro, R. (1998). The Drosophila Fab-7 chromosomal element coveys epigenetic inheritance during mitosis and meiosis. *Cell, 93*, 505–518.

Chan, C. S., Rastelli, L., & Pirrotta, V. (1994). A Polycomb response element in the Ubx gene that determines an epigenetically inherited state of repression. *The EMBO Journal*, *13*, 2553–2564.

Chanas, G., & Machat, F. (2005). Tissue specificity of hedgehog repression by the Polycomb group during *Drosophila melanogaster* development. *Mechanisms of Development*, *122*, 975–987.

Chang, Y. L., King, B., Lin, S. C., Kennison, J. A., & Huang, D. H. (2007). A double-bromodomain protein, FSH-S, activates the homeotic gene *Ultrabithorax* through a critical promoter-proximal region. *Molecular and Cellular Biology*, *15*, 5486–5489.

Cheng, Y., Kwon, D. Y., Arai, A. L., Mucci, D., & Kassis, J. A. (2012). P-element homing is facilitated by engrailed polycomb-group response elements in Drosophila melanogaster. *PLoS One*, *7*, e30437.

Cuddapah, S., Roh, T.-Y., Cui, K., Jose, C. C., Fuller, M. T., Zhao, K., et al. (2012). A novel human polycomb binding site acts as a functional polycomb response element in *Drosophila*. *PLoS One*, *7*, e36365.

Cunningham, M. D., Brown, J. L., & Kassis, J. A. (2010). Characterization of the Polycomb group response elements of the *Drosophila melanogaster invected* locus. *Molecular and Cellular Biology*, *30*, 820–828.

Cunningham, M. D., Gause, M., Cheng, Y., Noyes, A., Dorsett, D., Kennison, J. A., et al. (2012). Wapl antagonizes cohesin binding and promotes Polycomb-group silencing in Drosophila. *Development*, *139*, 4172–4179.

Czermin, B., Melfi, R., McCabe, D., Seitz, V., Imhof, A., & Pirrotta, V. (2002). Drosophila enhancer of Zeste/ESC complexes have a histone H3 methyltransferase activity that marks chromosomal Polycomb sites. *Cell*, *111*, 185–196.

Deaton, A., & Bird, A. (2011). CpG islands and the regulation of transcription. *Genes & Development*, *25*, 1010–1022.

Decoville, M., Giacomello, E., Leng, M., & Locker, D. (2001). DSP1, an HMG-like protein, is involved in the regulation of homeotic genes. *Genetics*, *157*, 237–244.

Dejardin, J., Rappailles, A., Cuvier, O., Grimaud, C., Decoville, M., Locker, D., et al. (2005). Recruitment of Drosophila Polycomb group proteins to chromatin by DSP1. *Nature*, *434*, 533–538.

Delest, A., Sexton, T., & Cavalli, G. (2012). Polycomb: A paradigm for genome organization from one to three dimensions. *Current Opinion in Cell Biology*, *24*, 405–414.

DeVido, S. K., Kwon, D., Brown, J. L., & Kassis, J. A. (2008). The role of Polycomb-group response elements in regulation of *engrailed* transcription in Drosophila. *Development*, *135*, 669–676.

Dorsett, D. (2011). Cohesin: Genomic insights into controlling gene transcription and development. *Current Opinion in Genetics and Development*, *21*, 199–206.

Drews, D., Klar, M., Dame, C., & Bräuer, A. U. (2009). Developmental expression profile of the $\gamma\gamma 2$ gene in mice. *BMC Developmental Biology*, *9*, 45.

Farkas, G., Gausz, J., Galloni, M., Reuter, G., Gyurkovics, H., & Karch, F. (1994). The Trithorax-like gene encodes the Drosophila GAGA factor. *Nature*, *371*, 806–808.

Fiedler, T., & Rehmsmeier, M. (2006). jPREdictor: A versatile tool for the prediction of cis-regulatory elements. *Nucleic Acids Research*, *34*, W546–W550.

Fritsch, C., Beuchle, D., & Müller, J. (2003). Molecular and genetic analysis of the Polycomb group gene sex combs extra/Ring in Drosophila. *Mechanisms of Development*, *120*, 949–954.

Fritsch, C., Brown, J. L., Kassis, J. A., & Müller, J. (1999). The DNA-binding Polycomb group protein Pleiohomeotic mediates silencing of a *Drosophila* homeotic gene. *Development*, *126*, 3905–3913.

Fujioka, M., Emi-Sarker, Y., Yusibova, G. L., Goto, T., & Jaynes, J. B. (1999). Analysis of an even-skipped rescue transgene reveals both composite and discrete neuronal and early blastoderm enhancers, and multi-stripe positioning by gap gene repressor gradients. *Development*, *126*, 2527–2538.

Fujioka, M., Wu, X., & Jaynes, J. B. (2009). A chromatin insulator mediates transgene homing and very long-range enhancer-promoter communication. *Development*, *136*, 3077–3087.

Fujioka, M., Yusibova, G. L., Zhou, J., & Jaynes, J. B. (2008). The DNA-binding Polycomb-group protein Pleiohomeotic maintains both active and repressed transcriptional states through a single site. *Development*, *135*, 4131–4139.

Gambetta, M. C., Oktaba, K., & Müller, J. (2009). Essential role of the glycosyltransferase Sxc/Ogt in Polycomb repression. *Science*, *325*, 93–96.

Garcia, E., Marcos-Gutierrez, C., del Mar Lorente, M., Moreno, J. C., & Vidal, M. (1999). RYBP, a new repressor protein that interacts with components of the mammalian Polycomb complex and with the transcription factor YY1. *The EMBO Journal*, *18*, 3404–3418.

Geini, R. S., & Hendzel, M. J. (2009). Polycomb group protein gene silencing, non-coding RNA, stem cells and cancer. *Biochemistry and Cell Biology*, *87*, 283–306.

Gindhart, J. G., & Kaufman, T. C. (1995). Identification of Polycomb and Trithorax Group responsive elements in the regulatory region of the *Drosophila* Homeotic Gene *Sex combs reduced*. *Genetics*, *139*, 797–814.

Goldberg, M. L., Colvin, R. A., & Mellin, A. F. (1989). The Drosophila zeste locus is non-essential. *Genetics*, *123*, 145–155.

Gordon, S., Akopyan, G., Garban, H., & Bonavida, B. (2006). Transcription factor YY1: Structure, function, and therapeutic implications in cancer biology. *Oncogene*, *25*, 1125–1142.

Grau, D. J., Chapman, B. A., Garlick, J. D., Borowsky, M., Francis, N. J., & Kingston, R. E. (2011). Compaction of chromatin by diverse Polycomb group proteins requires localized regions of high charge. *Genes and Development*, *25*, 2210–2221.

Grimm, C., Matos, R., Ly-Hartig, N., Stuerwald, U., Lindner, D., Rybin, V., et al. (2009). Molecular recognition of histone lysine methylation by the polycomb group repressor dSfmbt. *The EMBO Journal*, *28*, 1965–1977.

Groth, A. C., Fish, M., Nusse, R., & Calos, M. P. (2004). Construction of transgenic Drosophila by using the site-specific integrase from phage phi-C31. *Genetics*, *166*, 1775–1782.

Gruzdeva, N., Kyrchanova, O., Parshikov, A., Kullyev, A., & Georgiev, P. (2005). The Mcp element from the bithorax complex contains an insulator that is capable of pair-wise interactions and can facilitate enhancer-promoter communication. *Molecular and Cellular Biology*, *25*, 3682–3789.

Gustavson, E., Goldsborough, A. S., Ali, Z., & Kornberg, T. B. (1996). The Drosophila engrailed and invected genes: Partners in regulation, expression, and function. *Genetics*, *142*, 893–906.

Gutiérrez, L., Oktaba, K., Scheuermann, J. C., Gambetta, M. C., Ly-Hartig, N., & Müller, J. (2012). The role of the histone H2A ubiquitinase Sce in Polycomb repression. *Development*, *139*, 117–127.

Hagstrom, K., Müller, M., & Schedl, P. (1997). A *Polycomb* and GAGA dependent silencer adjoins the *Fab-7* boundary in the *Drosophila* bithorax complex. *Genetics*, *146*, 1365–1380.

Hama, C., Ali, Z., & Kornberg, T. B. (1990). Region-specific recombination and expression are directed by portions of the Drosophila engrailed promoter. *Genes and Development*, *4*, 1079–1093.

Hauenschild, A., Ringrose, L., Altmutter, C., Paro, R., & Rehmsmeier, M. (2008). Evolutionary plasticity of polycomb/trithorax response elements in Drosophila species. *PLoS Biology*, *28*, e261.

Horard, B., Tatout, C., Poux, S., & Pirrotta, V. (2000). Structure of a Polycomb response element and in vitro binding of Polycomb group complexes containing GAGA factor. *Molecular and Cellular Biology, 20*, 3187–3197.

Huang, D.-W., & Chang, Y.-L. (2004). Isolation and characterization of CHRASCH, a Polycomb-containing silencing complex. *Methods in Enzymology, 377*, 267–282.

Huang, D.-W., Chang, Y.-L., Yang, C.-C., Pan, I.-C., & King, B. (2002). *pipsqueak* encodes a factor essential for sequence-specific targeting of a polycomb group protein complex. *Molecular and Cellular Biology, 22*, 6261–6271.

Hur, M. W., Laney, J. D., Jeon, S. H., Ali, J., & Biggin, M. D. (2002). Zeste maintains repression of *Ubx* transgenes: Support for a new model of Polycomb repression. *Development, 129*, 1339–1344.

Jeon, Y., Sarma, K., & Lee, J. T. (2012). New and Xisting regulatory mechanisms of X inactivation. *Current Opinion in Genetics and Development, 22*, 62–71.

Kaczynski, J., Cook, T., & Urrutia, R. (2003). Sp1- and Kruppel-like transcription factors. *Genome Biology, 4*, 206.

Kassis, J. A. (1994). Unusual properties of regulatory DNA from the Drosophila *engrailed* gene: Three "pairing-sensitive" sites within a 1.6-kb region. *Genetics, 136*, 1025–1038.

Kassis, J. A. (2002). Pairing-sensitive silencing, Polycomb group response elements, and transposon homing in *Drosophila*. *Advances in Genetics, 46*, 421–438.

Kassis, J. A., Vansickle, E. P., & Sensabaugh, S. M. (1991). A fragment of *engrailed* regulatory DNA can mediate transvection of the *white* gene in Drosophila. *Genetics, 128*, 751–761.

Kapoun, A. M., & Kaufman, T. C. (1995). Regulatory regions of the homeotic gene proboscipedia are sensitive to chromosomal pairing. *Genetics, 140*, 643–658.

Kennison, J. A. (1995). The Polycomb and trithorax group proteins of Drosophila: Trans-regulators of homeotic gene function. *Annual Review of Genetics, 29*, 289–303.

Kerppola, T. K. (2009). Polycomb group complexes—Many combinations, many functions. *Trends in Cell Biology, 19*, 692–704.

Kharchenko, P. V., Alekseyenko, A. A., Schwartz, Y. B., Minoda, A., Riddle, N. C., Ernst, J., et al. (2011). Comprehensive analysis of the chromatin landscape in *Drosophila melanogaster*. *Nature, 471*, 480–485.

Kingsley, C., & Winoto, A. (1992). Cloning of GT box-binding proteins: A novel Sp1 multigene family regulating T-cell receptor gene expression. *Molecular and Cellular Biology, 12*, 4251–4261.

Klymenko, T., Papp, B., Fischle, W., Kocher, T., Schelder, M., Fritsch, C., et al. (2006). A Polycomb group protein complex with sequence-specific DNA-binding and selective methyl-lysine-binding activities. *Genes and Development, 20*, 1110–1122.

Kozma, G., Bender, W., & Sipos, L. (2008). Replacement of a Drosophila Polycomb response element core, and in situ analysis of its motifs. *Molecular Genetics and Genomics, 9*, 595–603.

Kwon, D., Mucci, D., Langlais, K., Americo, J. L., DeVido, S. K., Cheng, Y., et al. (2009). Enhancer-promoter communication at the *Drosophila engrailed* locus. *Development, 136*, 3067–3075.

Kwong, C., Adryan, B., Bell, I., Meadows, L., Russell, S., Manak, J. R., et al. (2008). Stability and dynamics of Polycomb target sites in Drosophila development. *PLoS Genetics, 4*, e1000178.

Lagarou, A., Mohd-sarip, A., Moshkin, Y. M., Chalkley, G. E., Bezatarosti, K., Demmers, J. A., et al. (2008). dKDM2 couples histone H2A ubiquitination to histone H3 demethylation during Polycomb group silencing. *Genes and Development, 22*, 2799–2810.

Lanzuolo, C., Roure, V., Dekker, J., Bantignes, F., & Orlando, V. (2007). Polycomb response elements mediate the formation of chromosome higher-order structures in the bithorax complex. *Nature Cell Biology, 9*, 1167–1174.

Lavigne, M., Francis, N. J., King, I. F., & Kingston, R. E. (2004). Propagation of silencing: Recruitment and repression of naïve chromatin in trans by polycomb repressed chromatin. *Molecular Cell, 13*, 415–425.

Lee, J. S., Calvin, K. M., & Shi, Y. (1993). Evidence for physical interaction between the zinc-finger transcription factors YY1 and Sp1. *Proceedings of the National Academy of Sciences, 90*, 6145–6149.

Lehmann, M., Siegmund, T., Lintermann, K., & Korge, G. (1998). The Pipsqueak protein of *Drosophila melanogaster* binds to GAGA sequences through a novel DNA-binding domain. *The Journal of Biological Chemistry, 273*, 28504–28509.

Lorente, M., Pérez, C., Sánchez, C., Donohoe, M., Shi, Y., & Vidal, M. (2006). Homeotic transformations of the axial skeleton of YY1 mutant mice and genetic interaction with the Polycomb group gene Ring1/Ring1A. *Mechanisms of Development, 123*, 312–320.

Lynch, M. D., Smith, A. J. H., De Gobbi, M., Fienley, M., Hughes, J. R., Vernimmen, D., et al. (2012). An interspecies analysis reveals a key role for undermethylated CpG dinucleotides in vertebrate Polycomb complex recruitment. *The EMBO Journal, 31*, 317–329.

Maeda, R. K., & Karch, F. (2011). Gene expression in time and space: Additive vs hierarchical organization of *cis*-regulatory regions. *Current Opinion in Genetics and Development, 21*, 187–193.

Mahmoudi, T., Zuijderduijn, L. M. P., Mohd-Sarip, A., & Verrijzer, C. P. (2003). GAGA facilitates binding of Pleiohomeotic to a chromatinized Polycomb response element. *Nucleic Acids Research, 31*, 4147–4156.

Matharu, N. K., Hussain, T., Sankaranarayanan, R., & Mishra, R. K. (2010). Vertebrate Homologue of *Drosophila* GAGA Factor. *Journal of Molecular Biology, 400*, 434–447.

Maurange, C., & Paro, R. (2002). A cellular memory module conveys epigenetic inheritance of hedgehog expression during *Drosophila* wing imaginal disc development. *Genes and Development, 16*, 2672–2683.

McConnell, B. B., & Yang, V. W. (2010). Mammalian Kruppel-like factors in health and diseases. *Physiological Reviews, 90*, 1337–1381.

McQuilton, P., St Pierre, S. E., Thurmon, J., & The FlyBase Consortium, (2012). FlyBase 101—The basics of navigating FlyBase. *Nucleic Acids Research, 40*, D706–D714.

Mendenhall, E. M., Koche, R. P., Truong, T., Zhou, V. W., Issac, B., Chi, A. S., et al. (2010). GC-Rich Sequence Elements recruit PRC2 in Mammalian ES Cells. *PLOS Genetics, 6*, e1001244.

Mishra, R. K., Mihaly, J., Barges, S., Spierer, A., Karch, F., Hagstrom, K., et al. (2001). The iab-7 Polycomb response element maps to a nucleosome-free region of chromatin and requires both GAGA and Pleiohomeotic for silencing activity. *Molecular and Cellular Biology, 21*, 1311–1318.

Moazed, D., & O'Farrell, P. H. (1992). Maintenance of the *engrailed* expression pattern by Polycomb group genes in Drosophila. *Development, 116*, 805–810.

Mohd-Sarip, A., Cleard, F., Mishra, R. K., Karch, F., & Verrijzer, C. P. (2005). Synergistic recognition of an epigenetic DNA element by Pleiohomeotic and a Polycomb core complex. *Genes and Development, 19*, 1755–1760.

Mohd-Sarip, A., Van der Knaap, J. A., Wyman, C., Kanaar, R., Schedl, P., & Verrijzer, C. P. (2006). Architecture of a polycomb nucleoprotein complex. *Molecular Cell, 24*, 91–100.

Mohd-Sarip, A., Venturini, F., Chalkley, G. E., & Verrijzer, C. P. (2002). Pleiohomeotic can link Polycomb to DNA and mediate transcriptional repression. *Molecular and Cellular Biology, 22*, 7473–7483.

Müller, J., & Bienz, M. (1991). Long range repression conferring boundaries of Ultrabithorax expression in the Drosophila embryo. *The EMBO Journal, 10*, 3147–3155.

Müller, J., Hart, C. M., Francis, N. J., Vargas, M. L., Sengupta, A., Wild, B., et al. (2002). Histone methyl transferase activity of a Drosophila Polycomb group repressor complex. *Cell*, *111*, 197–208.

Müller, J., & Kassis, J. A. (2006). Polycomb response elements and targeting of Polycomb group proteins in *Drosophila*. *Current Opinion in Genetics and Development*, *16*, 476–484.

Müller, J., & Verrijzer, P. (2009). Biochemical mechanisms of gene regulation by polycomb group protein complexes. *Current Opinion in Genetics and Development*, *19*, 150–158.

Négre, N., Hennetin, J., Sun, L. V., Lavrov, S., Bellis, M., White, K. P., et al. (2006). Chromosomal distribution of PcG proteins during *Drosophila* development. *PLoS Biology*, *4*, 170.

Nekrasov, M., Klymenko, T., Fraterman, S., Papp, B., Oktaba, K., Köcher, T., et al. (2007). Pcl-PRC2 is needed to generate high levels of H3-K27 trimethylation at Polycomb target genes. *The EMBO Journal*, *26*, 4078–4088.

Nguyen, N., Zhang, X., Olashaw, N., & Seto, E. (2004). Molecular cloning and functional characterization of the transcription factor YY2. *The Journal of Biological Chemistry*, *279*, 25927–25934.

Noyes, A., Stefaniuk, C., Cheng, Y., Kennison, J. A., & Kassis, J. A. (2011). Modulation of the activity of a Polycomb-group response element in Drosophila by a mutation in the transcriptional activator Woc. *G3 (Bethesda)*, *1*, 471–478.

Oktaba, K., Gutiérrez, L., Gagneur, J., Girardot, C., Sengupta, A., Furlong, E. E. M., et al. (2008). Dynamic regulation of Polycomb group protein complexes controls pattern formation and the cell cycle in *Drosophila*. *Developmental Cell*, *15*, 877–889.

Okulski, H., Druck, B., Bhalerao, S., & Ringrose, L. (2011). Quantitative analysis of Polycomb response elements (PREs) at identical genomic locations distinguishes contributions of PRE sequence and genomic environment. *Epigenetics and Chromatin*, *4*, 1–16.

Pagans, S., Ortiz-Lombardia, M., Esinas, M. L., Bernues, J., & Azorin, F. (2002). The *Drosophila* transcription factor *tramtrack* (TTK) Interacts with *Trithorax-like* (GAGA) and represses GAGA-mediated activation. *Nucleic Acids Research*, *30*, 4406–4413.

Papp, B., & Müller, J. (2006). Histone trimethylation and the maintenance of transcriptional ON and OFF states by trxG and PcG proteins. *Genes and Development*, *20*, 2041–2054.

Park, S. Y., Schwartz, Y. B., Kahn, T. G., Asker, D., & Pirrotta, V. (2012). Regulation of Polycomb group genes Psc and Su(z)2 in *Drosophila melanogaster*. *Mechanisms of Development*, *128*, 536–547.

Pérez-Lluch, S., Cuartero, S., Azorin, F., & Espinås, M. L. (2008). Characterization of new regulatory elements within the Drosophila bithorax complex. *Nucleic Acids Research*, *36*, 6926–6933.

Pirrotta, V. (1988). Vectors for P-mediated transformation in Drosophila. *Biotechnology*, *10*, 437–456.

Pirrotta, V., & Li, H.-B. (2012). A view of nuclear Polycomb bodies. *Current Opinion in Genetics and Development*, *22*, 101–109.

Poux, S., Kostic, C., & Pirrotta, V. (1996). Hunchback-independent silencing of late Ubx enhancers by a Polycomb Group Response Element. *The EMBO Journal*, *15*, 4713–4722.

Rank, G., Prestel, M., & Paro, R. (2002). Transcription through intergenic chromosomal memory elements of the Drosophila bithorax complex correlates with an epigenetic switch. *Molecular and Cellular Biology*, *22*, 8026–8034.

Rappailles, A., Decoville, M., & Locker, D. (2005). DSP1, a *Drosophila* HMG protein, is involved in spatiotemporal expression of the homeotic gene *Sex combs reduced*. *Biology of the Cell*, *97*, 779–785.

Ringrose, L., & Paro, R. (2004). Epigenetic regulation of cellular memory by the Polycomb and trithorax group proteins. *Annual Review of Genetics*, *38*, 413–443.

Ringrose, L., & Paro, R. (2007). Polycomb/Trithorax response elements and epigenetic memory of cell identity. *Development*, *134*, 223–232.

Ringrose, L., Rehmsmeier, M., Dura, J. M., & Paro, R. (2003). Genome-wide prediction of Polycomb/Trithorax response elements in *Drosophila melanogaster*. *Developmental Cell*, *5*, 759–771.

Rubin, G. M., & Spradling, A. C. (1982). Genetic transformation of Drosophila with transposable element vectors. *Science*, *218*, 348–353.

Salvaing, J., Decoville, M., Mouchel-Vielh, E., Bussiere, M., Daulny, A., Boldyreva, L., et al. (2006). Corto and DSP1 interact and bind to a maintenance element of the *Scr Hox* gene: Understanding the role of *Enhancers of trithorax and Polycomb*. *BMC Biology*, *4*, 9.

Saurin, A. J., Shao, Z., Erdjument-Bromage, H., Tempst, P., & Kingston, R. E. (2001). A Drosophila Polycomb group complex includes Zeste and dTAFII proteins. *Nature*, *412*, 655–660.

Savla, U., Benes, J., Zhang, J., & Jones, R. S. (2008). Recruitment of Drosophila Polycomb-group proteins by Polycomblike, a component of a novel protein complex in larvae. *Development*, *135*, 813–817.

Scheuermann, J. C., de Ayala Alonso, A. G., Oktaba, K., Ly-Hartig, N., McGinty, R. K., Fraterman, S., et al. (2010). Histone H2A deubiquitinase activity of the Polycomb repressive complex PR-DUB. *Nature*, *465*, 243–247.

Schmitt, S., Prestel, M., & Paro, R. (2005). Intergenic transcription through a polycomb group response element counteracts silencing. *Genes and Development*, *19*, 697–708.

Schuettengruber, B., & Cavalli, G. (2009). Recruitment of polycomb group complexes and their role in the dynamic regulation of cell fate choice. *Development*, *136*, 3531–3542.

Schuettengruber, B., & Cavalli, G. (2010). The DUBle life of polycomb complexes. *Developmental Cell*, *18*, 878–880.

Schuettengruber, B., Ganapathi, M., Leblanc, B., Portoso, M., Jaschek, R., Tolhuis, B., et al. (2009). Functional anatomy of Polycomb and Trithorax chromatin landscapes in *Drosophila* embryos. *PLoS Biology*, *7*, 0001–0018.

Schwartz, Y. B., Kahn, T. G., Nix, D. A., Li, X. Y., Bourgon, R., Biggin, M., et al. (2006). Genome-wide analysis of Polycomb targets in *Drosophila melanogaster*. *Nature Genetics*, *38*, 700–705.

Schwartz, Y. B., Kahn, T. G., Stenberg, P., Ohno, K., Bourgon, R., & Pirrotta, V. (2010). Alternative epigenetic chromatin states of Polycomb target genes. *PLoS Genetics*, *6*, e1000805.

Schwendemann, A., & Lehman, M. (2002). Pipsqueak and GAGA factor act in concert as partners at homeotic and many other loci. *Proceedings of the National Academy of Sciences of the United States of America*, *99*, 12883–12888.

Seto, E., Lewis, B., & Shenk, T. (1993). Interaction between transcription factors Sp1 and YY1. *Nature*, *365*, 462–464.

Shao, Z., Raible, F., Mollaaghababa, R., Guyon, J. R., Wu, C. T., Bender, W., et al. (1999). Stabilization of chromatin structure by PRC1, a Polycomb complex. *Cell*, *98*, 37–46.

Shimell, M. J., Peterson, A. J., Burr, J., Simon, J. A., & O'Connor, M. B. (2000). Functional analysis of repressor binding sites in the *iab-2* regulatory region of the *abdominal-A* homeotic gene. *Developmental Biology*, *218*, 38–52.

Shintomi, D., & Hirano, T. (2009). Releasing cohesin from chromosome arms in early mitosis: Opposing actions of Wapl-Pds5 and Sgo1. *Genes and Development*, *23*, 2224–2236.

Siegel, V., Jongens, T. A., Jan, L. Y., & Jan, Y. N. (1993). Pipsqueak, an early acting member of the posterior group of genes, affects vasa level and germ cell-somatic cell interaction in the developing egg chamber. *Development*, *119*, 1187–1202.

Siegmund, T., & Lehmann, M. (2002). The *Drosophila* Pipsqueak protein defines a new family of helix-turn-helix DNA binding proteins. *Development Genes and Evolution*, *212*, 152–157.

Simon, J., Chiang, A., Bender, W., Shimell, M. J., & O'Connor, M. (1993). Elements of the Drosophila bithorax complex that mediate repression by Polycomb group products. *Developmental Biology*, *158*, 131–144.

Simon, J. A., & Kingston, R. E. (2009). Mechanisms of Polycomb gene silencing: Knowns and unknowns. *Nature Reviews Molecular Cell Biology*, *10*, 697–708.

Sing, A., Pannell, D., Karaiakakis, A., Sturgeon, K., Djabali, M., Ellis, J., et al. (2009). A vertebrate Polycomb response element governs segmentation of the posterior hindbrain. *Cell*, *138*, 885–897.

Smouse, D., Goodman, C., Mahowald, A., & Perrimon, N. (1988). Polyhomeotic: A gene required for the embryonic development of axon pathways in the central nervous system of Drosophila. *Genes and Development*, *2*, 830–842.

Sparmann, A., & van Lohuizen, M. (2006). Polycomb silencers control cell fate, development and cancer. *Nature Reviews Cancer*, *6*, 846–856.

Stros, M. (2010). HMGβ proteins: Interactions with DNA and chromosomes. *Biochimica et Biophysica Acta*, *1799*, 101–113.

Strübbe, G., Popp, C., Schmidt, A., Pauli, A., Ringrose, L., Beisel, C., et al. (2011). Polycomb purification by in vivo biotinylation tagging reveals cohesin and Trithorax group proteins as interaction partners. *PNAS*, *108*, 5572–5577.

Strutt, H., Cavalli, G., & Paro, R. (1997). Co-localization of Polycomb protein and GAGA factor on regulatory elements responsible for the maintenance of homeotic gene expression. *The EMBO Journal*, *16*, 3621–3632.

Suske, G., Bruford, E., & Philipsen, S. (2005). Mammalian SP/KLF transcription factors: Bring in the family. *Genomics*, *85*, 551–556.

Taillenbourg, E., & Dura, J. M. (1999). A novel mechanism for P element homing in Drosophila. *Proceedings of the National Academy of Sciences*, *96*, 6856–6861.

Tolhuis, B., de Wit, E., Muijrers, I., Teunissen, H., Talhout, W., van Steensel, B., et al. (2006). Genome-wide profiling of PRC1 and PRC2 Polycomb chromatin binding in *Drosophila* melanogaster. *Nature Genetics*, *38*, 694–699.

Tuckfield, A., Clouston, D. R., Wilanowski, T. M., Zhao, L. L., Cunningham, J. M., & Jane, S. M. (2002). Binding of the RING polycomb proteins to specific target genes in a complex with the grainyhead-like family of developmental transcription factors. *Molecular and Cellular Biology*, *22*, 1936–1946.

Vazquez, J., Muller, M., Pirrotta, V., & Sedat, J. W. (2006). The Mcp Element Mediates Stable Long-Range Chromosome-Chromosome Interactions in *Drosophila*. *Molecular Biology of the Cell*, *17*, 2158–2165.

Venkatesan, K., McManus, H. R., Mello, C. C., Smith, T. F., & Hansen, U. (2003). Functional conservation between family members of an ancient duplicated transcription factor family. *Nucleic Acids Research*, *31*, 4304–4316.

Vernimmen, D., Lynch, M. D., De Gobbi, M., Garrick, D., Sharpe, J. A., Sloane-Stanley, J. A., et al. (2011). Polycomb eviction as a new distant enhancer function. *Genes and Development*, *25*, 1583–1588.

Wang, L., Brown, J. L., Cao, R., Zhang, Y., Kassis, J. A., & Jones, R. S. (2004). Hierarchical recruitment of polycomb group silencing complexes. *Molecular Cell*, *14*, 637–646.

Wang, C.-C., Chen, J. J. W., & Yang, P.-C. (2006). Multifunctional transcription factor YY1: A therapeutic target in human cancer. *Expert Opinion on Therapeutic Targets*, *10*, 253–266.

Wang, L., Jahren, N., Miller, E. L., Ketel, C. S., Mallin, D. R., & Simon, J. A. (2010). Comparative analysis of chromatin binding by sex Comb on Midleg (SCM) and other Polycomb group repressors at a *Drosophila Hox* gene. *Molecular and Cellular Biology*, *30*, 2584–2593.

Woo, C. J., Kharchenko, P. V., Daheron, L., Park, P. J., & Kingston, R. E. (2010). A region of the human HOXD Cluster that confers polycomb-group responsiveness. *Cell*, *140*, 99–110.

Zeng, J., Kirk, B. D., Gou, Y., Wang, Q., & Ma, J. (2012). Genome-wide polycomb target prediction in *Drosophila melanogaster*. *Nucleic Acids Research*, *40*, 5848–5863.

INDEX

Note: Page numbers followed by "*f*" indicate figures, and "*t*" indicate tables.

A

Adhesins
  a-agglutinin, 61–62
  A. fumigatus proteins, 63
  α-agglutinin and Sag1p, 61–62
  Als genes, 62–63
  C. albicans, 62–63
  C. neoformans proteins, 63–64
  flocculation, 61–62
  Hwp1p, 63
ARE. *See* Asymptotic relative efficiency
    (ARE)
Aspergillus fumigatus cell wall, 67–68
Asymptotic relative efficiency (ARE)
  association, 15
  definition, 14
  heritability, 14–15
  MLEs, 14
  polygenic model, 20–21
  Wald statistic, 14

B

Bgl2p/Scw4p/Scw10p/Scw11p family, β-
    1,3-glucanases, 58–59

C

Candida albicans cell wall, 67
Cell wall biogenesis, 55–56
Cell wall glycoproteins
  "cytosolic" proteins, 50
  GPI anchoring, 50–52
  GPI- and non-GPI anchored proteins,
    49–50
  integral cell wall proteins, 40*t*, 49
  N-linked and O-linked mannans and
    galactomannans, 52–55
Cell wall integrity (CWI) response,
    64–65
Cell wall proteins, 65–66
Chitin and chitosan
  mutational analysis, 43
  nikkomycins and polyoxins, 42–43

trafficking and activation aspects,
    39–42
  UDP-*N*-acetylglucosamine, 39
Chitinases, 60
Complex human pedigrees
  analysis, GRKs, 23–26
  ARE, 14–15
  eigensimplification approach, 26
  ELRT (*see* Expected likelihood ratio test
    (ELRT))
  EVD, 3–4
  GWA, 2–3
  linear mixed model, 3
  NGS and WGS, 2–3
  polygenic model (*see* Polygenic model)
  power, 12–13
  VCs models (*see* Variance components
    (VCs) models)
Components, fungal cell walls
  cell wall glycoproteins (*see* Cell wall
    glycoproteins)
  chitin and chitosan, 39–43
  α-1,3-glucans, 47
  β-1,3-glucans, 43–45
  β-1,6-glucans, 45–46
  melanins, 47–49
  mixed β-1,3-/β-1,4-glucans, 45
Crh/Utr family, β-1,3-glucanases, 58
Cryptococcus neoformans cell wall, 68

D

Drosophila
  PcG proteins, 84–85
  PREs in
    DNA binding proteins and sequence
      motifs, 95–102
    flanking regulatory DNA, 92–94
    mini-*white* silencing, 88–92
    PRE maintenance assays, 86–88
    transcriptional activation/repression,
      94–95
Dsp1, 101

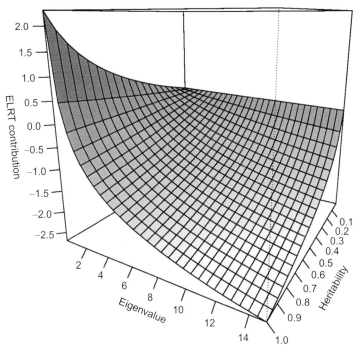

Chapter 1, Figure 1.2. (See Page 10 of this volume).

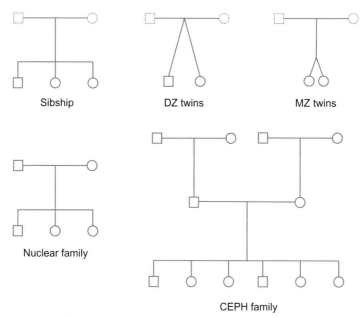

Sibship   DZ twins   MZ twins

Nuclear family

CEPH family

Chapter 1, Figure 1.3. (See Page 16 of this volume).

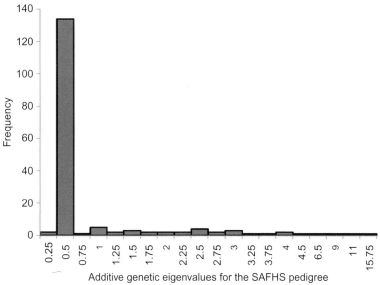

Chapter 1, Figure 1.5. (See Page 18 of this volume).

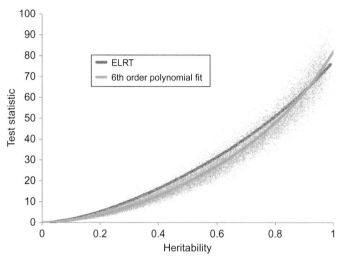

Chapter 1, Figure 1.6. (See Page 19 of this volume).

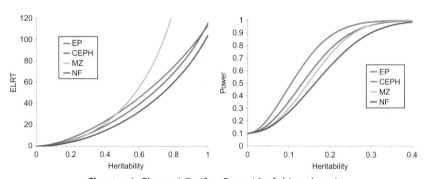

Chapter 1, Figure 1.7. (See Page 19 of this volume).

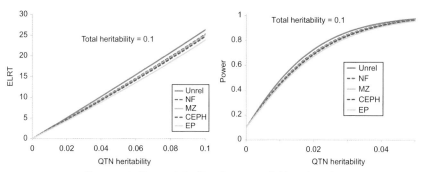

Chapter 1, Figure 1.8. (See Page 20 of this volume).

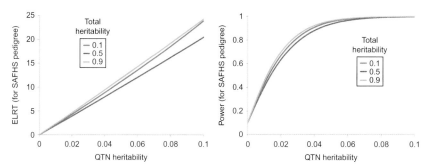

Chapter 1, Figure 1.9. (See Page 20 of this volume).

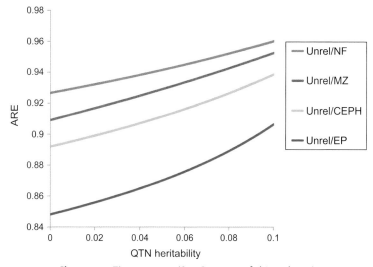

Chapter 1, Figure 1.10. (See Page 21 of this volume).

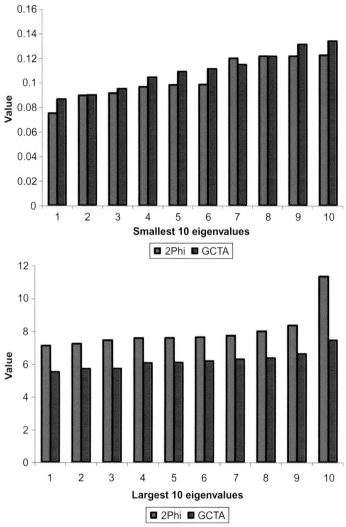

Chapter 1, Figure 1.11. (See Page 25 of this volume).

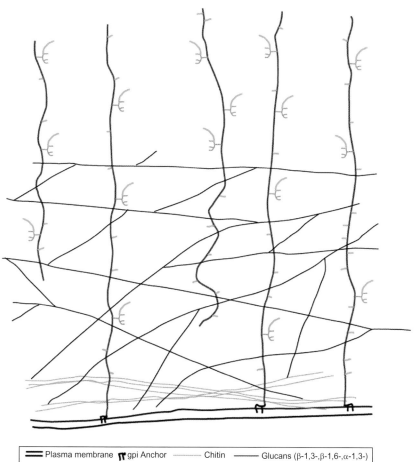

══ Plasma membrane	⑉ gpi Anchor	----------- Chitin	———— Glucans (β-1,3-,β-1,6-,α-1,3-)
━━ Protein	⊂E N-linked oligosaccharide	——— Galactomannan or mannan	

Chapter 2, Figure 2.1. (See Page 37 of this volume).

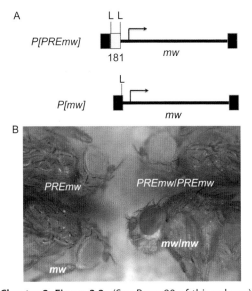

Chapter 3, Figure 3.2. (See Page 90 of this volume).